AF357343

Fiber Optic Sensors and Fiber Lasers

Fiber Optic Sensors and Fiber Lasers

Editor

Min Yong Jeon

MDPI • Basel • Beijing • Wuhan • Barcelona • Belgrade • Manchester • Tokyo • Cluj • Tianjin

Editor
Min Yong Jeon
Department of Physics
Chungnam National University
Daejeon
Korea, South

Editorial Office
MDPI
St. Alban-Anlage 66
4052 Basel, Switzerland

This is a reprint of articles from the Special Issue published online in the open access journal *Sensors* (ISSN 1424-8220) (available at: www.mdpi.com/journal/sensors/special_issues/fiber_optic_sensors_lasers).

For citation purposes, cite each article independently as indicated on the article page online and as indicated below:

LastName, A.A.; LastName, B.B.; LastName, C.C. Article Title. *Journal Name* **Year**, *Volume Number*, Page Range.

ISBN 978-3-0365-1518-2 (Hbk)
ISBN 978-3-0365-1517-5 (PDF)

Contents

About the Editor

Min Yong Jeon

Min Yong Jeon is a full professor of Chungnam National University of Korea. For over 30 years, he has been a significant and active researcher in fiber optics and photonics. In 1994, he received his Ph. D in Physics from the Korea Advanced Institute of Science and Technology (KAIST). From 1994 to 2001, he worked on optical fiber lasers, photonic packet switching, OTDM technologies, and wavelength conversions in a basic research laboratory of the Electronics and Telecommunications Research Institute (ETRI), Daejeon, Korea as a senior member. From 2001 to 2003, he worked on all-optical label swapping and optical label switching system in the all-optical network laboratory of University of California at Davis as a Research Scientist. Since 2003, he has joined the department of Physics at Chungnam National university of Korea as a professor, where he has been continuing his research on fiber-optics and photonics.

Preface to "Fiber Optic Sensors and Fiber Lasers"

The optical fiber industry is emerging from the market for selling simple accessories using optical fiber to the new optical-IT convergence sensor market combined with high value-added smart industries such as the bio industry. Among them, fiber optic sensors and fiber lasers are growing faster and more accurately by utilizing fiber optics in various fields such as shipbuilding, construction, energy, military, railway, security, and medical.

This Special Issue aims to present novel and innovative applications of sensors and devices based on fiber optic sensors and fiber lasers and covers a wide range of applications of optical sensors. In this Special Issue, original research articles, as well as reviews, have been published.

Two review papers related to distributed Brillouin fiber sensors and ultrafast fiber lasers and eight research papers related to lasers and sensors have been published. In particular, I would like to thank Ms. Xanthe Du for her active help in publishing this Special Issue.

Min Yong Jeon
Editor

Review

Recent Progress in Distributed Brillouin Sensors Based on Few-Mode Optical Fibers

Yong Hyun Kim and Kwang Yong Song *

Department of Physics, Chung-Ang University, Seoul 06974, Korea; kim.yonghyun@kt.com
* Correspondence: songky@cau.ac.kr; Tel.: +82-2-820-5834

Abstract: Brillouin scattering is a dominant inelastic scattering observed in optical fibers, where the energy and momentum transfer between photons and acoustic phonons takes place. Narrowband reflection (or gain and loss) spectra appear in the spontaneous (or stimulated) Brillouin scattering, and their linear dependence of the spectral shift on ambient temperature and strain variations is the operation principle of distributed Brillouin sensors, which have been developed for several decades. In few-mode optical fibers (FMF's) where higher-order spatial modes are guided in addition to the fundamental mode, two different optical modes can be coupled by the process of stimulated Brillouin scattering (SBS), as observed in the phenomena called intermodal SBS (two photons + one acoustic phonon) and intermodal Brillouin dynamic grating (four photons + one acoustic phonon; BDG). These intermodal scattering processes show unique reflection (or gain and loss) spectra depending on the spatial mode structure of FMF, which are useful not only for the direct measurement of polarization and modal birefringence in the fiber, but also for the measurement of environmental variables like strain, temperature, and pressure affecting the birefringence. In this paper, we present a technical review on recent development of distributed Brillouin sensors on the platform of FMF's.

Keywords: few-mode fiber; fiber optic sensors; Brillouin scattering; distributed measurement

Citation: Kim, Y.H.; Song, K.Y. Recent Progress in Distributed Brillouin Sensors Based on Few-Mode Optical Fibers. *Sensors* **2021**, *21*, 2168. https://doi.org/10.3390/s21062168

Academic Editor: Min Yong Jeon

Received: 27 February 2021
Accepted: 17 March 2021
Published: 19 March 2021

Publisher's Note: MDPI stays neutral with regard to jurisdictional claims in published maps and institutional affiliations.

1. Introduction

Since the first experimental demonstration of the temperature and strain measurement based on Brillouin scattering in a single-mode fiber (SMF) in late 1980s [1,2], the key performances of distributed sensors such as sensing range, spatial resolution, measurement time, and accuracy have been significantly improved over the past decades, thanks to the advances on optical methodologies, signal processing techniques, and optoelectronic devices [3–7]. Brillouin sensors using an SMF can provide a long measurement distance over 100 km in obtaining local Brillouin gain spectrum (BGS), however, accurate discrimination between the effects of temperature and strain variation is still challenging although a few approaches have been proposed [7,8]. In optical communications, a few-mode fiber (FMF), together with a multi-core fiber (MCF), have received considerable attention as a potential platform for space division multiplexing (SDM) [9,10], attracted by the demand for increasing transmission capacity. In optical fiber sensors as well as optical communications, different spatial modes in an FMF can find unique applications such as simultaneous measurement of multiple variables, and this has been the main target in the development of Brillouin sensors based on an FMF.

As most of commercially available fiber-optic components are made of SMF, the handling and analysis of signals in a specific higher-order mode of FMF requires an efficient and selective mode converter that couples the fundamental mode and the target higher-order mode. Several active and passive mode converters have been developed in the form of directional coupler, long-period fiber grating, and phase plate [11–15]. Among them a mode selective coupler (MSC) is a directional coupler that provides mode coupling between the fundamental and higher-order modes with low-loss, broad bandwidth, and high mode-purity, which is suitable for the FMF-based Brillouin sensor system [16,17]. For the methodology of distributed sensing, an optical time-domain or correlation-domain approach is used to obtain the BGS of intermodal SBS, similar to the cases of SMF-based Brillouin sensors [18,19]. The Brillouin dynamic grating (BDG), in which the acoustic phonons generated by the SBS of pump waves are used for the reflection of probe wave, can be demonstrated and characterized in various configurations using the FMF between different optical modes [20,21].

In this paper, recent experimental results of the FMF-based distributed Brillouin sensors are revisited. The spectral characteristics of the Brillouin gain spectrum (BGS) of the intra- and intermodal SBS using the LP_{01} and LP_{11} modes in the e-core TMF are presented in Section 2, where additional multi-peak features of the intermodal SBS in a circular-core FMF are also presented. Brillouin frequency measurement using optical time-domain analysis on the SBS of the e-core TMF is provided in Section 3 with the loss characterization of LP_{11} mode according to the bending radius and direction. Analysis on the double peak BGS uniquely shown in the intramodal SBS of LP_{11} mode is described in Section 4, where the optical correlation-domain analysis is applied with differential measurement schemes. The optical time-domain measurements of the intermodal BDG spectra with two LP modes of the e-core TMF are demonstrated in Section 5, where the temperature and strain coefficients of BDG frequency for different pairs of pump-probe modes are determined. The result of discriminative sensing of temperature and strain distribution is presented by applying optical time-domain reflectometry of the BDG spectrum in the e-core TMF.

2. Brillouin Scattering in Few-Mode Optical Fibers

Local temperature and strain variation can be measured through the spectral shift of the BGS in the distributed Brillouin sensor [22,23]. The reflection spectrum of spontaneous Brillouin scattering of the fundamental LP_{01} mode in an SMF is generally composed of multiple Lorentzian curves with the full width at half maximum (FWHM) of about 20–30 MHz [22–24], each of which generally corresponds to different acoustic modes in the fiber. When the configuration of Brillouin amplifier is adopted, only a single Lorentzian curve is dominantly obtained as BGS by SBS. The frequency offset between the pump and probe waves with the maximum Brillouin gain on the probe is called Brillouin frequency (ν_B), which is given by [23]:

$$\nu_B = \frac{2n_{eff}V_a}{\lambda} \tag{1}$$

where n_{eff}, V_a, and λ are effective refractive index (ERI), acoustic velocity, and wavelength of the light source, respectively. Since the V_a of acoustic mode is a function of acoustic frequency, the ν_B is theoretically determined as a common solution between optical and acoustic dispersion relations. At the wavelength of 1550 nm, ERI of 1.45, and acoustic velocity of ~5900 m/s, the ν_B is calculated to be ~11 GHz. For accurate measurement of the BGS and ν_B of optical fiber, one needs a narrowband (<1 MHz) light source, a photo detector (PD), microwave devices, and electro-optic modulators for the control or measurement of frequency offset between two optical waves. The localization of sensing positions in distributed Brillouin sensors additionally requires high-speed electronics like a short-pulse generator, a high-speed PD, a high-speed data acquisition system for time-domain schemes, and accurate frequency- or phase-modulation devices with a lock-in amplifier for the correlation-domain systems. Although suffered by the system complexity and high cost, Brillouin sensors can provide truly distributed measurement of strain and temperature when compared to FBG-based point sensors. The longer sensing range (over 100 km) and the higher spatial resolution (order of cm or sub-cm) are advantageous features of Brillouin sensors when compared to Raman-scattering-based distributed temperature sensor (DTS). Dependencies of ν_B on temperature and strain variations are the characteristics of test fiber itself, and the coefficients of conventional SMF's are known as ~1 MHz/°C and ~0.05 MHz/$\mu\varepsilon$, respectively [23]. For distributed sensing of temperature and strain by ν_B, the measurement of reference values should be performed first, and at least two independent sets of measurement results with different coefficients are needed to discriminate the effects of temperature and strain. In the intermodal SBS of FMF, a higher-order spatial mode like the LP_{11} mode is used, which has anti-symmetric electric field (E-field) distribution in contrast to the symmetric distribution of the fundamental mode. Since the acoustic wave is generated by the electrostriction caused by the moving interference pattern of the E-fields of optical waves, the intermodal SBS between the LP_{01} and LP_{11} modes is necessarily intervened by anti-symmetric acoustic modes showing unique BGS with Lorentzian shape.

In a circular-core fiber, the LP_{11} mode is an approximate mode composed of almost degenerate TM_{01}, TE_{01}, and HE_{21} modes. This feature results in unstable orientation of the intensity lobe in the propagation along the fiber. When the core is elliptical, on the contrary, the LP_{11} mode splits into two groups, i.e., $LP_{11}{}^{odd}$ and $LP_{11}{}^{even}$ modes, with considerably different ERI's and well-defined and stable intensity patterns. When the size and ellipticity of core are properly designed, the $LP_{11}{}^{odd}$ mode is cut-off while the $LP_{11}{}^{even}$ mode is still guided. This fiber is called e-core TMF, supporting two stable spatial modes, i.e., the LP_{01} and the $LP_{11}{}^{even}$ modes [25,26]. The e-core TMF is useful for several applications such as fiber sensors, tunable filters, and fiber lasers [27,28] thanks to the stable lobe orientation in the LP_{11} mode. The e-core TMF used in our work has the index difference Δ of 0.6% and the core radius of 5.4 μm × 3.6 μm, which guides only the LP_{01} and the LP_{11} even modes with the LP_{11} odd mode cut-off at 1550 nm. The polarization birefringence Δn (for LP_{01} mode) is about 3.5×10^{-5} [16].

For the measurement of BGS with the e-core TMF, a polished-type mode selective coupler (MSC) was used for the selective launch and retrieval of each mode [16]. The MSC is fabricated using a pair of SMF and e-core TMF, where the ERI of the LP_{11} mode in the e-core TMF is set equal to that of the LP_{01} mode in the SMF. Figure 1 shows the operation of MSC. The coupling efficiency between the LP_{01} mode of SMF and the LP_{11} mode of TMF was ~80%, and the purity of generated mode in the TMF was 23 dB, respectively. The inset shows the far fields from the output end of the e-core TMF measured after selective launching of each mode by the MSC, which reflects the clear images of LP_{01} and LP_{11} modes [16].

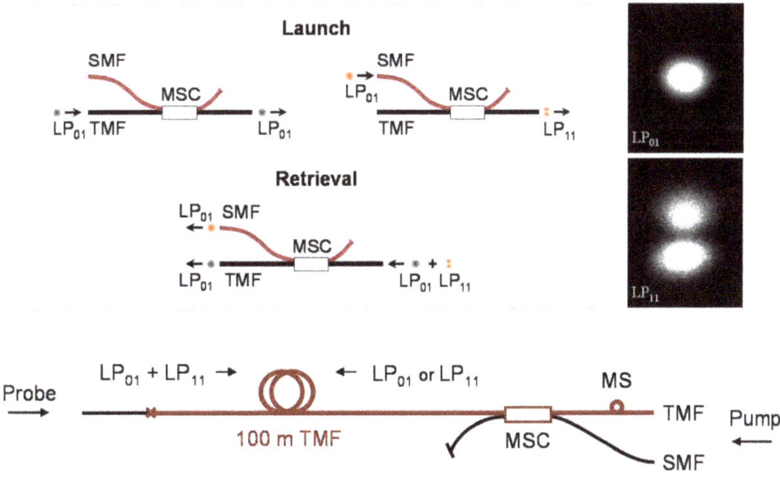

Figure 1. Operation of the MSC for launch and retrieval of each mode (upper), and the connection of fiber under test (FUT) and MSC for the BGS measurement of the e-core TMF. Reprinted with permission from Ref. [16] ©The Optical Society.

Figure 2a–d are the BGS of four possible pairs of pump-probe modes of LP_{01}-LP_{01}, LP_{01}-LP_{11}, LP_{11}-LP_{01}, and LP_{11}-LP_{11} modes, respectively, in the e-core TMF, showing the

FWHM as indicated [16]. It is interesting to see that the BGS of the intramodal SBS of LP_{01} mode and the intermodal SBS between LP_{11} and LP_{01} modes show a single Lorentzian peak with ~30 MHz of FWHM, while the BGS of the intramodal SBS of LP_{11} mode has two dominant peaks of comparable size. Although a single peak appears in both Figure 2a,b, it should be noted that the intramodal SBS of LP_{01} mode is intervened by a symmetric acoustic mode while the intermodal SBS between LP_{01} and LP_{11} modes is by an anti-symmetric acoustic mode, due to the symmetry of E-field distribution of interacting optical modes. Such a difference can be supported by the observed ~10% difference in the FWHM, which reflects the difference in the lifetime of acoustic phonon in two scatterings. It is also thought that the multiple peaks of the intramodal SBS of LP_{11} mode are attributed to higher-order symmetric acoustic modes, and the simulation results in [16] show that the measured v_B differences in Figure 2 match well with simulations with discrepancy of less than ±3 MHz. In the FMFs, the phase-matching condition of the Brillouin scattering presented in Equation (1) is rewritten as:

$$v_B = \left(\frac{n_i}{\lambda_i} + \frac{n_j}{\lambda_j} \right) V_a \approx \frac{V_a}{\lambda} \left(n_i + n_j \right) \tag{2}$$

where $n_{i(j)}$ and $\lambda_{i(j)}$ are the ERI and wavelength of *ith* (*jth*) mode, respectively. It should be noted that the acoustic velocity (V_a) changes according to the acoustic mode.

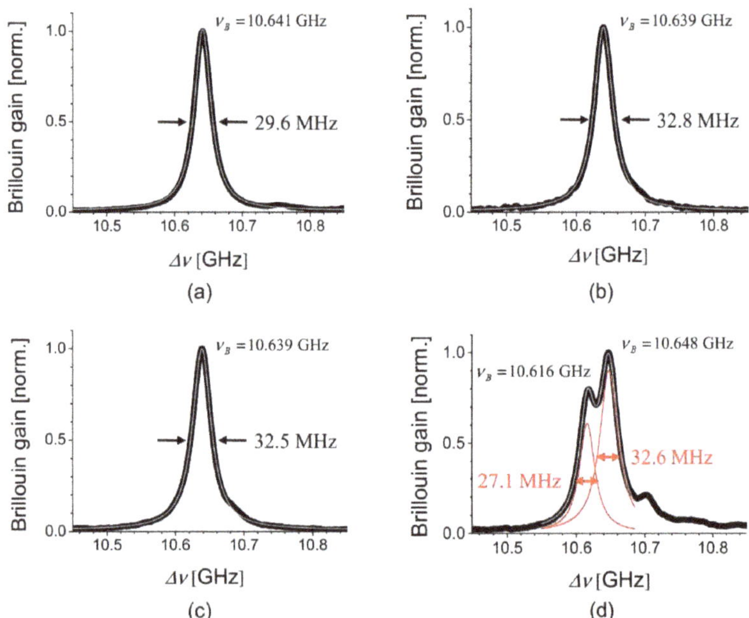

Figure 2. BGS of different pairs of pump-probe modes in the e-core TMF [16]: (**a**) LP_{01}-LP_{01}; (**b**) LP_{01}-LP_{11}; (**c**) LP_{11}-LP_{01}; and (**d**) LP_{11}-LP_{11} modes. Reprinted with permission from Ref. [16] ©The Optical Society.

The Brillouin gain as a function of pump power for the four different pairs is plotted in Figure 3, where the length of the e-core TMF was 100 m. When compared to the case of intramodal SBS of fundamental mode, the relative magnitude of gain of intermodal SBS (or intramodal SBS of LP_{11} mode) is 0.58 (or 0.47), both of which are large enough to be used for sensing applications [16].

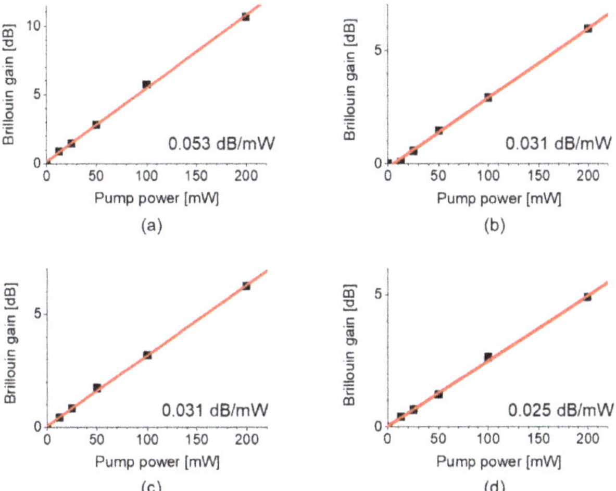

Figure 3. Brillouin gain as a function of pump power for the pump-probe pairs of (**a**) LP_{01}-LP_{01}, (**b**) LP_{01}-LP_{11}, (**c**) LP_{11}-LP_{01}, and (**d**) LP_{11}-LP_{11} modes, plotted with a line fit for each. Reprinted with permission from Ref. [16] ©The Optical Society.

When the fiber was changed from the e-core TMF to circular-core four-mode fiber (FoMF), more peaks were observed in the BGS, so the envelope of BGS no longer remains as a single Lorentzian in most cases of pump-probe modes [17]. Figure 4a–d are the BGS of intramodal SBS in the FoMF using the LP_{01}, LP_{11}, LP_{21}, or LP_{02} mode for both pump and probe waves. In all four cases, four large and small peaks were found in the BGS, and in particular the BGS of LP_{21} mode shows two dominant peaks with comparable gain while the rest commonly have three small peaks with a single dominant peak. It is thought that the intramodal SBS in the FoMF is intervened by a series of symmetric acoustic modes, which is supported by the fact that the amount of frequency separation between each of the four peaks is common in all the intramodal BGS. If the order of the acoustic mode is numbered to increase from small to large frequency, those numbers for dominant acoustic mode are not identical in those four cases.

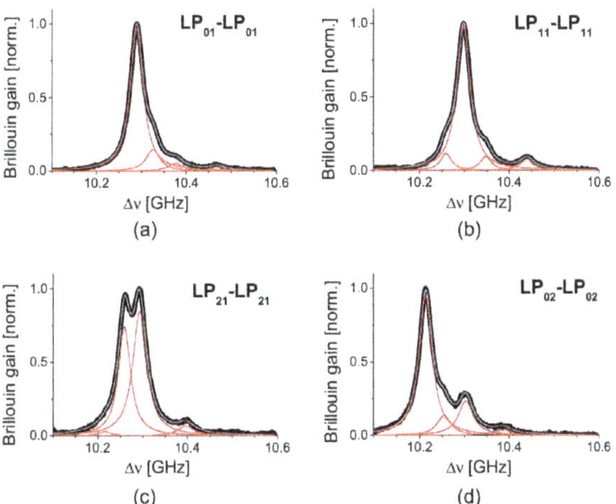

Figure 4. The BGS of the intramodal SBS of (**a**) LP_{01}, (**b**) LP_{11}, (**c**) LP_{21}, and (**d**) LP_{02} modes in the FoMF. Reprinted with permission from Ref. [17] ©The Optical Society.

Multiple peaks are also observed in the BGS of intermodal SBS in the FoMF for the pump-probe pair of LP_{01}-LP_{11}, LP_{01}-LP_{21}, LP_{01}-LP_{02}, LP_{11}-LP_{21}, LP_{11}-LP_{02}, and LP_{21}-LP_{02} as presented in Figure 5a–f [17]. Unlike the intramodal SBS, the BGS of the intermodal SBS can be fitted with different numbers (from 1 to 4) of Lorentzian curves.

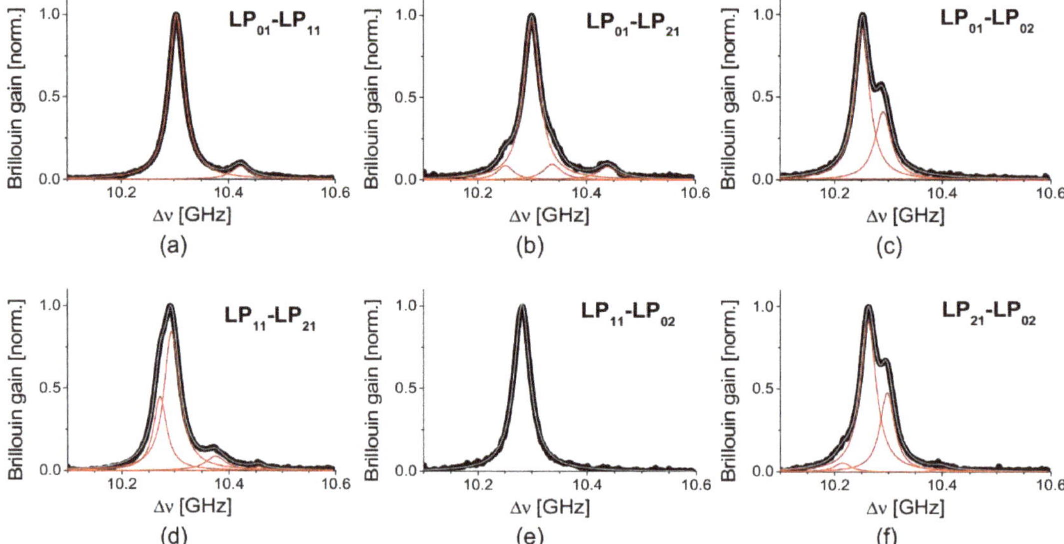

Figure 5. The BGS of the intermodal SBS for the pump-probe pairs of (**a**) LP_{01}-LP_{11}, (**b**) LP_{01}-LP_{21}, (**c**) LP_{01}-LP_{02}, (**d**) LP_{11}-LP_{21}, (**e**) LP_{11}-LP_{02}, and (**f**) LP_{21}-LP_{02} modes in the FoMF, fitted with different numbers (1–3) of Lorentzian curves for each. Reprinted with permission from Ref. [17] ©The Optical Society.

The relative magnitude of the gain coefficients of the intramodal (or intermodal) SBS in the FoMF were reported within 35–55% (or 14–45%) of that of the LP_{01} mode [17]. To build a distributed Brillouin sensor system using a FoMF, the pump-probe mode pair should be carefully selected to secure a clear and large signal free from possible intermodal interference. The difference of the SBS threshold from that of the fundamental mode in the FoMF was measured to be 0.6, 2.7, and 2.8 dB for the LP_{11}, LP_{21}, and LP_{02} mode, respectively [17].

Additionally, the measurement of the BGS of a circular-core two-mode fiber was reported, applying free-space mode coupling [29], and the results show single Lorentzian gain curves for both intra- and intermodal SBS. It is thought that further investigation is needed for the quantitative analysis on the acoustic modes involved in the SBS of FMF's.

3. Optical Time-Domain Brillouin Sensor Using a Few-Mode Fiber

Optical time-domain reflectometry and analysis have been the most popular way for implementing distributed Brillouin sensors, thanks to the simple configuration and intuitive signal structure compared to frequency and correlation domain methods [1–3,30]. The time of flight of the pump pulse directly corresponds to the sensing position, and the spatial resolution of the system is determined only by the duration of pulse. In long-distance applications of FMF's, the propagation loss of a higher-order mode is expected to be significantly larger than that of the fundamental mode, and its distribution, as well as the distribution of other ambient variables, is measurable for each mode by applying Brillouin optical time domain analysis (BOTDA).

3.1. Bending Loss Characteristic of the LP_{11} Mode in the E-Core TMF

The optical loss induced by an unwanted macro-bending could become a potential weakness either in communication or a sensor based on higher-order spatial modes. Some periodic notches are commonly detected in transmitted optical power as the bending radius is reduced, which is known as the resonances of the whispering gallery modes at the core/cladding boundary [31]. The bending loss characteristics of the LP_{11} modes in the e-core TMF were measured with polyethylene mandrel and rotation stages. Figure 6 shows the measurement result according to the bending radius with three different bending directions relative to the orientation of core ellipse when half-round bending is applied [32].

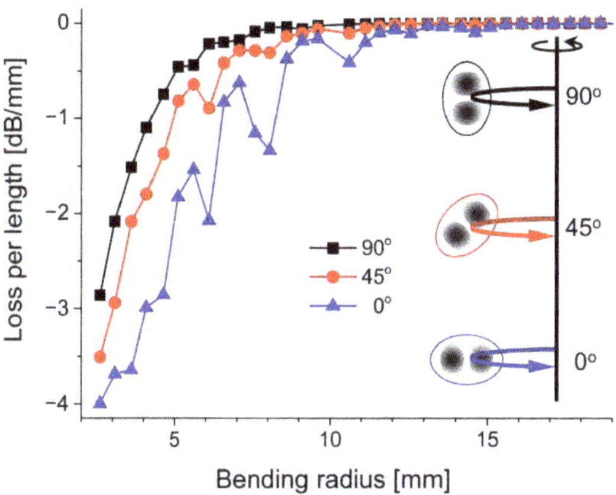

Figure 6. Loss of the LP_{11} mode per unit length in the e-core TMF according to bending radius [32].

The mandrel was made as a co-axial cylinder having a diameter varying from 5.2–46.2 mm with 1 mm steps. Three different angles of $0°$, $45°$, and $90°$ between the plane of the bending and major axis of the elliptic core were applied to check the loss characteristics. When the plane of bending is parallel to the major axis of core (i.e., $0°$), one can observe some periodic notches were observed in the transmitted power with overall the largest of loss among three cases. Meanwhile, the notches disappear when the plane of bending becomes orthogonal (i.e., $90°$) with overall the smallest loss. Such orientation-dependent loss is thought to originate from the π-phase difference between the E-fields of two intensity lobes, which may suppress the coupling to whispering gallery modes by the offset of two contributions at $90°$.

3.2. Distributed Brillouin Sensor for Bending Loss Detection

Brillouin optical time-domain analysis (BOTDA) and reflectometry (BOTDR) are simple and intuitive ways for distributed Brillouin sensing. The experimental setup of the BOTDA system based on the e-core TMF is presented in Figure 7 [32]. A distributed feedback laser diode (DFB-LD) with a center wavelength of 1550 nm was used as a light source, and the output was divided by into the pump and probe arms. The probe was generated by an electro-optic modulator (EOM) and a microwave generator (MWG), and the pump was modulated as a pulse with a duration of 20 ns corresponding to the spatial resolution of 2 m. The polarization of the pump was switched between two orthogonal states of polarization by a polarization switch (PSW) for acquiring the average value of Brillouin gain. The LP_{01} mode was launched as the probe to the fiber under test, a 65 m e-core TMF, through a mode stripper (MS), i.e., tight bend, and the LP_{01} and LP_{11} modes were selectively launched as the pump from the opposite end by using an MSC as depicted in the inset 'FUT'. The MSC was also used to selectively retrieve the LP_{01} and LP_{11} mode components of the probe for signal acquisition. A fiber Bragg grating (FBG) was used to filter out the anti-Stokes components of two sidebands.

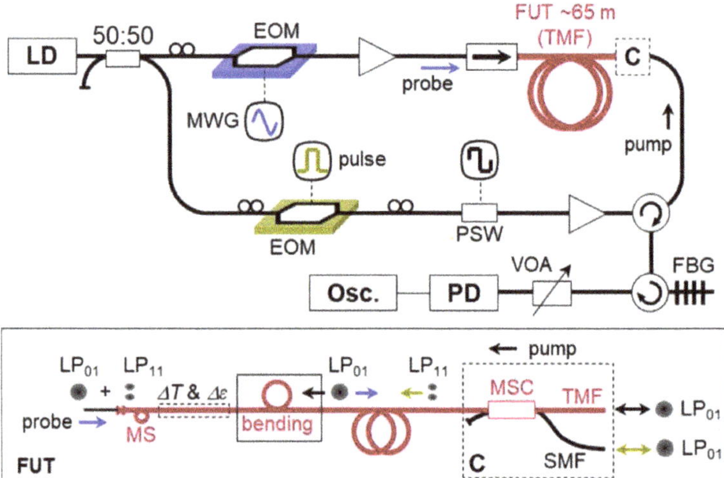

Figure 7. Setup of the BOTDA system based on the e-core TMF [32]: EOM, electro-optical modulator; MWG, microwave generator; PSW, polarization switch; FBG, fiber Bragg grating; VOA, variable optical attenuator; PD, photo detector; FUT, fiber under test; MSC, mode selective coupler.

The variation of ν_B with respect to temperature and strain change was measured from the single-peak BGS of the intra- and intermodal SBS, as seen in Figure 2. Examples of the measured BGS at the strain-applied section and the distribution map of ν_B along the fiber are plotted in Figure 8a,b, respectively, where variable strain with 400 με step was applied to the 2.5 m test section near the end of FUT. The average ν_B of the intermodal SBS in the FUT was ~10.642 GHz [32].

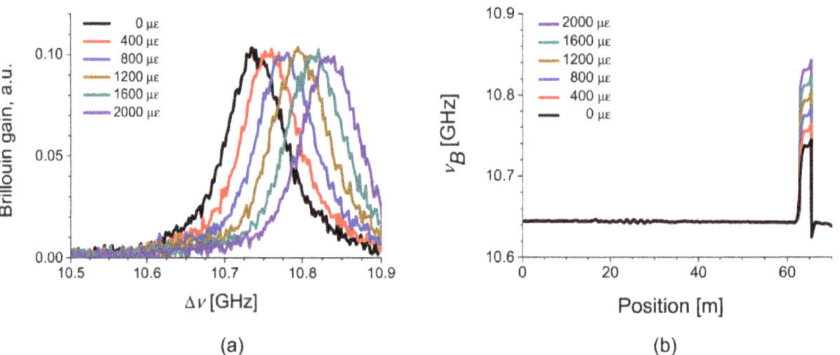

Figure 8. (**a**) BGS at the test section with different strains for the intermodal SBS. (**b**) Distribution map of ν_B with variable strains at the test section for the intermodal SBS [32].

The temperature (C_T) and strain (C_ε) coefficients of the ν_B for different pairs of pump-probe modes were measured by the experimental setup in Figure 7, and the results are presented in Figure 9a,b for the pump-probe pair of LP_{01}-LP_{01}, and Figure 9c,d for the LP_{11}-LP_{01}, respectively [32]. The measured coefficients of 0.047 MHz/με for strain and 1.07 (or 1.08) MHz/°C for temperature are similar to those of conventional SMF, and the distinction between the intra- and intermodal SBS is negligible, which indicates that it would not be possible to discriminate the temperature and strain by using those two pairs of the e-core TMF.

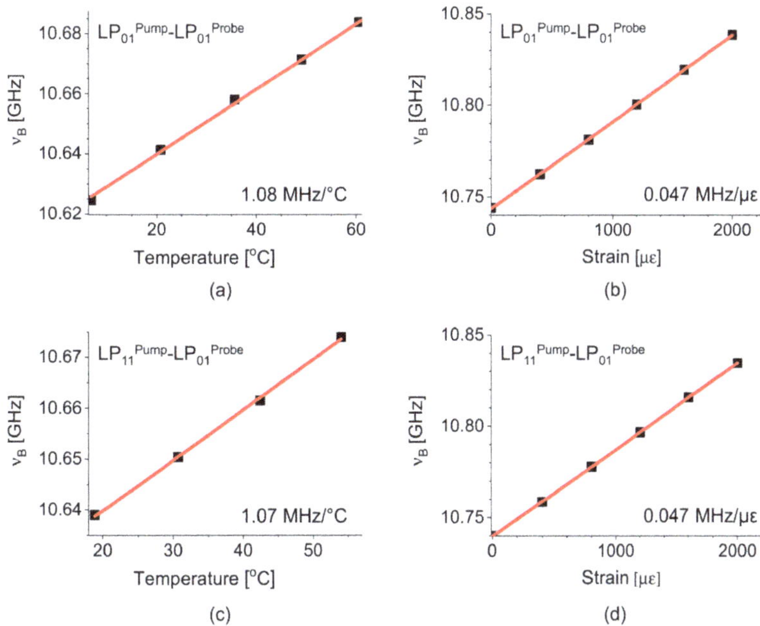

Figure 9. Temperature and strain dependence of the ν_B for (**a**,**b**) the intramodal SBS of LP_{01} mode and (**c**,**d**) the intermodal SBS between LP_{11} and LP_{01} modes, respectively, in the e-core TMF [32].

Although the diameter of unwanted macro-bending cannot be quantified due to the existence of resonance notches, the presence of the bending loss is detectable by measuring the gain reduction in the intermodal SBS. Two different bends with diameters of 1.4 and 1.5 cm were applied at the position of 36 and 56 m from the input port of the pulse for simultaneous measurement of the loss of LP_{11} mode and ν_B. The measurement results are plotted in Figure 10, where Figure 10a,b show the distribution maps of the BGS and ν_B of the intermodal SBS, and Figure 10c is the map of Brillouin gain normalized to its maximum for the intra- (black, LP_{01} mode) and intermodal (red) SBS, respectively. One can see the change of signal by two local bends, such as the small variation of ν_B in Figure 10b and the abrupt reduction of the Brillouin gain in Figure 10a,c. The shift of ν_B near the rear end of FUT is observed in Figure 10a,b by intentional temperature change from 25 to 54 °C [32].

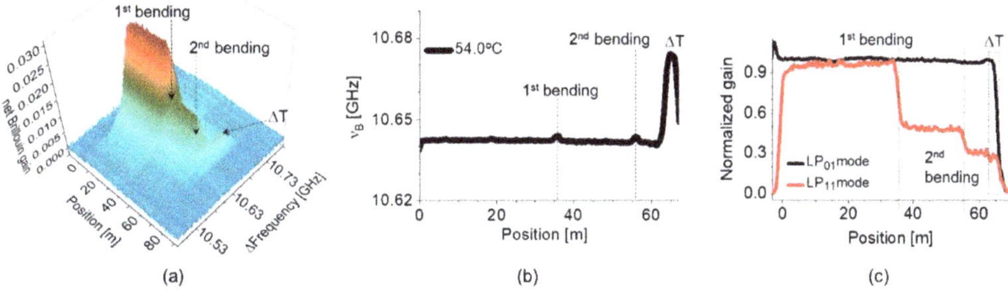

Figure 10. Simultaneous measurement of the loss of LP_{11} mode and ν_B based on the e-core TMF: (**a**) the distribution map of BGS of intermodal SBS, (**b**) the distribution map of ν_B of intermodal SBS, and (**c**) the distribution map of Brillouin gain normalized by the maximum for the intra- (black, LP_{01} mode) and intermodal (red) SBS [32].

Local BGS measured at different positions are shown in Figure 11a,b, respectively. Figure 11a is the BGS of intramodal (LP_{01} mode) SBS where the BGS in black, red, and green correspond to the positions of 35, 38, and 58 m after zero, one, and two turns of bending, respectively, at ambient temperature of 25.5 °C. The BGS in blue is the result at 65 m with the temperature of 18.2 °C, also after two turns of bending. One can see the shift of BGS while the Brillouin gains are maintained. Figure 11b is the BGS of intermodal SBS where the BGS in black, red, green, and blue represent the results of the same positions as those in Figure 11a except that the temperature was 30.8 °C for the result at 65 m (blue). In intermodal SBS, the gain reduction is clearly observed, which is measured to be 49% and 39%, respectively, by the first and second bending at different diameters (14.1 and 15.1 mm). It is notable that the peak gain of intermodal SBS (black) was about 59% of that of the SBS between LP_{01} modes, matching well with the result obtained by CW interaction of the pump and probe waves [16].

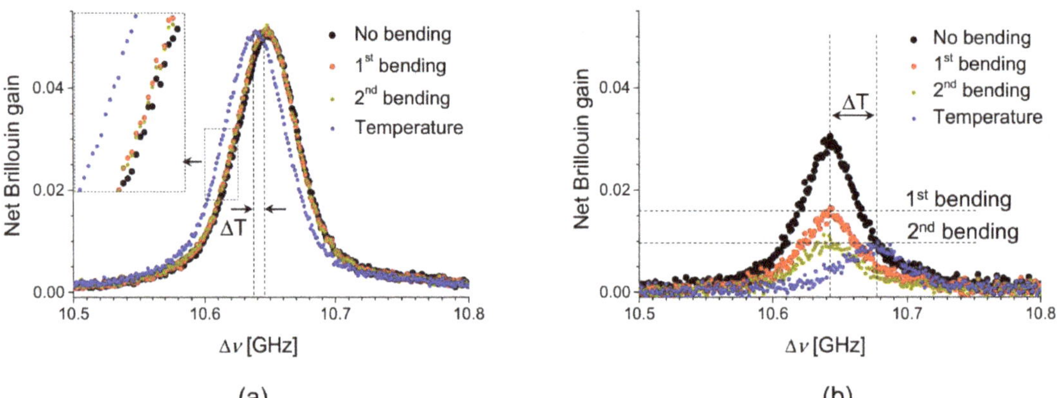

Figure 11. Local BGS at the different positions after bending of zero, one, and two turns and temperature variation after the second bending for (**a**) the intramodal (LP_{01} mode) and (**b**) the intermodal SBS [32].

It is worth mentioning that the ratio of Brillouin gain of the intermodal SBS to that of the intramodal (LP_{01} mode) SBS is 0.6 in a circular core TMF [29], which is quite similar to 0.58, the ratio of the e-core TMF [16]. Additionally, the strain coefficients of intra- (LP_{01} mode) and intermodal SBS (LP_{01} and LP_{11}) of the e-core TMF were almost the same as 470 MHz/%, while quite dissimilar values were reported for the intramodal SBS (592 MHz/% for LP_{01}, and 487 MHz/% for LP_{11} mode, respectively) of a circular-core FMF [15]. The BOTDA based on the circular-core FMF has been successfully applied for the discriminative measurement of strain and temperature, thanks to the large discrepancy in the strain coefficients [15]. It is notable that the measurement with a circular-core FMF shows better accuracy in discriminating the effects of two environmental variables than that with the e-core TMF. However, the difficulties in the development of optical components and the low stability of Brillouin interaction for the LP_{11} mode in the circular-core fiber, due to the unstable lobe orientation, are other important issues to be solved for its practical applications. The BOTDA system based on the FMF provides an advantage of simultaneous measurement of bending loss, strain and temperature variations, while the TMF-based OTDR system can measure only the loss distribution of the higher order mode [33].

4. Optical Correlation-Domain Brillouin Sensor Using a Few-Mode Fiber

There are three different pairs of pump-probe modes for the intra- and intermodal BGS in the e-core TMF, and the temperature and strain dependences of ν_B distribution for two pairs (LP_{01}-LP_{01} and LP_{11}-LP_{01}) were measured applying the BOTDA system. It is necessary to change the sensing scheme to investigate the temperature and strain dependence of the double-peak BGS of the intramodal SBS of LP_{11} mode in the e-core TMF, presented in Figure 2d, since the frequency separation (~32 MHz) between two peaks is too close [32].

4.1. Brillouin Optical Correlation-Domain Analysis with Differential Measurement Scheme

The FWHM of BGS is reported to be about 30 MHz at the wavelength of 1550 nm in a conventional SMF. In general, the BGS measured by an optical time-domain Brillouin sensor is the spectral convolution of the intrinsic Lorentzian BGS and the spectrum of pump wave. The spectral width of BGS measured by a BOTDA system is generally broader than the intrinsic spectral width (i.e., ~30 MHz) due to the finite spectral width of the pump pulse [34]. Figure 12a shows the BGS of the intramodal SBS of the LP_{11} mode, with a continuous wave (CW) pump (red) and a 20 ns-pulsed pump (black) [32]. It is seen that the double peak structure in the CW-based BGS completely disappears when the pulsed pump is applied. Therefore, one needs another approach to characterize the double-peak BGS of intramodal SBS of the LP_{11} mode.

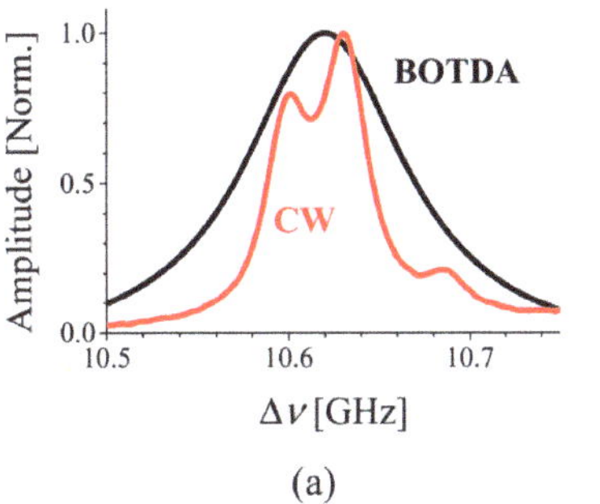

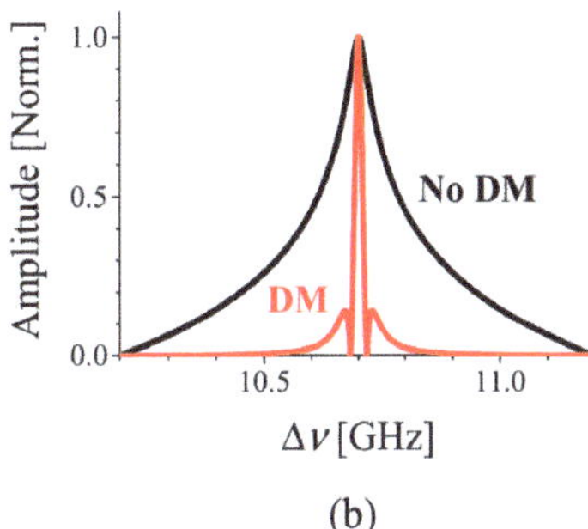

Figure 12. (**a**) Double-peak BGS of the LP_{11} mode measured by CW pump and probe waves (red) and by a BOTDA system with a 20 ns pulse as the pump (black), respectively. (**b**) Example of BGS of an SMF measured by a BOCDA using the ordinary configuration (black) and the DM with a 5 MHz phase modulation (red) [32].

The Brillouin optical correlation-domain analysis (BOCDA) has been developed for high-resolution distributed Brillouin sensing with CW pump and probe waves on the basis of synthesis of optical coherence function (SOCF) for localizing the sensing position [19]. Differential measurement (DM) is a modified lock-in detection technique that was introduced to the BOCDA system for improving the spatial resolution and dynamic range of distributed measurement [35]. The DM-BOCDA system is operated by applying on-off phase modulation to the pump to suppress the noise substructure in the BGS. An additional notable advantage of the DM-BOCDA is the acquisition of BGS with a spectral width narrower than the intrinsic Brillouin linewidth of 30 MHz. Figure 12b shows examples of BGS obtained by the ordinary (black) and the DM-based (red) BOCDA systems for an SMF, where the FWHM of signal obtained by the DM-BOCDA system is only about 17 MHz while that of ordinary BOCDA system is as large as 170 MHz.

4.2. BOCDA System Based on the Intramodal SBS of LP_{11} Mode

Figure 13 is the experimental setup of the DM-BOCDA based on the intramodal SBS of LP_{11} mode in the e-core TMF [32], which is a linearly configured system using a 1550 nm DFB-LD as the light source. The modulation frequency (f_m) and depth (Δf) were about 3 MHz and 1.65 GHz, respectively, which corresponds to the measurement range of 33 m and the nominal spatial resolution of 20 cm [19].

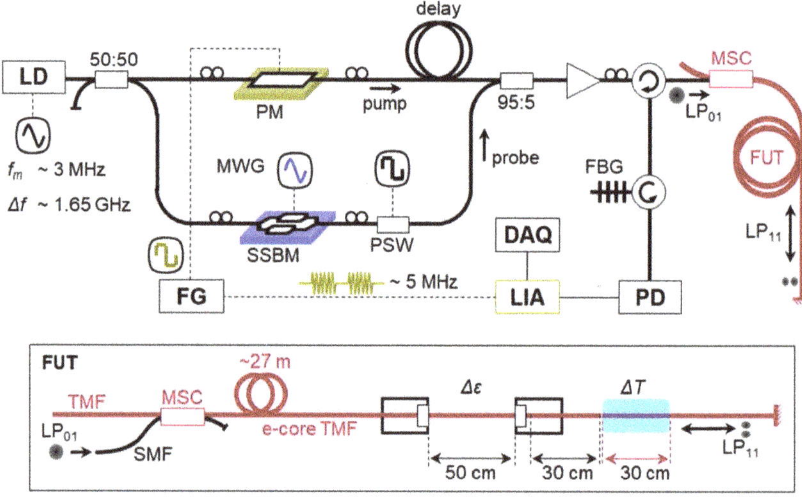

Figure 13. Experimental setup of the BOCDA using the e-core TMF [32]: PM, phase modulator; FG, function generator; SSBM, single sideband modulator; MWG, microwave generator; PSW, polarization switch; MSC, mode selective coupler; FUT, fiber under teste; FBG, fiber Bragg grating; PD, photo detector; LIA, lock-in amplifier; DAQ, data acquisition.

The output from DFB-LD was divided by a 3 dB coupler for the probe and pump waves, which were separately modulated by using a single sideband modulator (SSBM) and a phase modulator (PM), respectively. In particular, the frequency of the phase modulation of pump was set as 5 MHz that was periodically turned on and off at 91 kHz, and this on-off frequency was used as the reference frequency of a lock-in amplifier (LIA) for DM. The frequency offset of the probe was swept from 10.5 to 10.8 GHz for the acquisition of BGS. The pump and probe waves were combined and launched to an FUT (e-core TMF) from one end in the LP_{11} mode through an MSC. The intramodal SBS of the LP_{11} mode occurred between the outgoing pump and the incoming probe (i.e., 4% Fresnel reflection from the cleaved end of FUT). The probe component was selected by an FBG among the reflected waves, and it was received by a PD for lock-in detection.

The distribution of the BGS of the intramodal SBS of LP_{11} mode measured by the DM-BOCDA system is plotted in Figure 14a, and the zoomed view of the test sections (dashed box) near the end of FUT is shown in Figure 14b. One can observe the shift of double-peak BGS according to the variations of temperature (ΔT = 12.9 °C) and strain ($\Delta \varepsilon$ = 1000 µε) applied to the spans of 30 cm and 50 cm, respectively. Figure 14c,d show the examples of local BGS at the test section, where the double-peak structure and the shift of the BGS of intramodal SBS of LP_{11} mode are clearly seen with different strains (0 and 2500 µε) and temperatures (27.4 and 52.5 °C) applied.

Figure 15a–d show the strain and temperature dependence of ν_B for the intramodal SBS of LP_{11} mode, measured separately for each peak in the double-peak BGS, where we set the peak order according to the ν_B's from low to high. The strain and temperature coefficients of the first (second) peak were measured as 0.047 MHz/µε (for both), and 1.06 (1.05) MHz/°C, respectively. It is notable that the measured coefficients of the two peaks are very close to those of the intermodal and intramodal SBS of fundamental modes with the discrepancy less than 1%. Since the two peaks in the intramodal SBS of LP_{11} mode have nearly the same coefficients for ambient variables, it is expected that the intramodal SBS of LP_{11} mode can be also used for the BOTDA system, by ignoring the double-peak structure and applying single-peak fitting to the combined BGS.

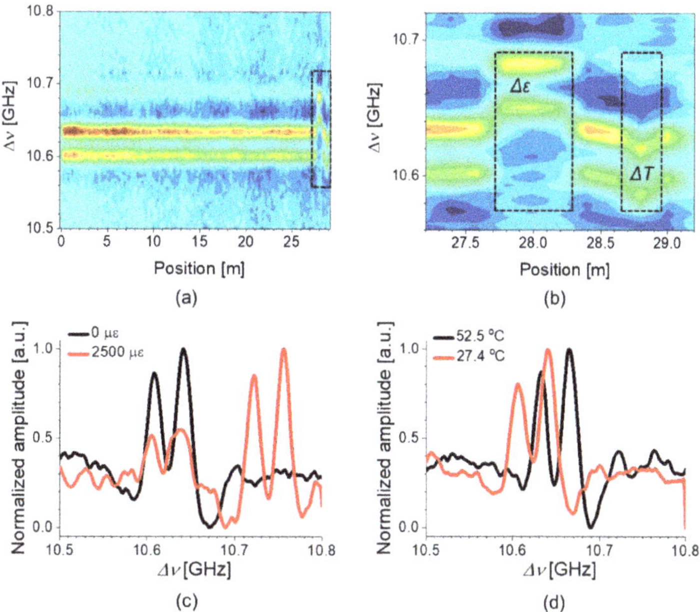

Figure 14. (**a**) Distribution map of the double-peak BGS measured by the DM-BOCDA system. (**b**) Zoomed view of the 2-m test section near the end of FUT. Local BGS of the intramodal SBS of LP_{11} mode at the position where (**c**) strain and (**d**) temperature variations were applied [32].

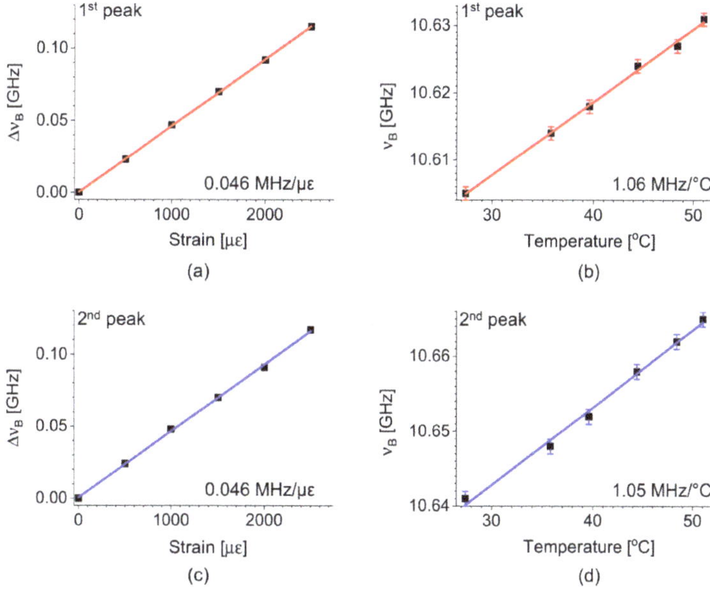

Figure 15. The shift of ν_B of the double-peak BGS in the intramodal SBS of LP_{11} mode, for the first peak according to (**a**) strain and (**b**) temperature, and for the second peak according to (**c**) strain and (**d**) temperature, respectively [32].

The BOCDA using FMF might be applicable to distributed sensing with a cm-order spatial resolution for discriminative measurement, if the relative amplitude or frequency separation of two peaks in the intramodal BGS of LP_{11} mode shows a significantly large discrepancy in the coefficients of two different physical variables. Another possible application is the development of linearly configured BOCDA system applying intermodal SBS, where selective reflection of each optical mode at the end of FUT could be used as the counter-propagating wave.

5. Brillouin Dynamic Grating Sensor Based on a Few-Mode Fiber

Brillouin dynamic grating (BDG) represents an acoustic phonon, which is generated in the process of SBS of optical waves called 'pump' and plays the role of a moving grating for another wave called 'probe' [20]. Several applications have been reported on the basis of BDG such as tunable delay lines, microwave filters, all-optical signal processing, and distributed sensors [36–44]. The BDG is typically implemented using a birefringent medium like a polarization maintaining fiber (PMF), and in the reflection spectrum of BDG (or BDG spectrum) the frequency offset between the pump and probe waves is called BDG frequency (v_D), which is a function of birefringence. The sensors based on BDG measure the ambient variables like strain, temperature, and pressure through the change of local birefringence [39–44]. When compared to ordinary Brillouin sensors, the BDG-based sensors can provide about 20 and 50 times better sensitivities for the strain and temperature measurement with conventional PMF's [39,40], and distributed sensors of hydrostatic pressure have also been reported as applying the BDG, where the sensitivity is at least 60 times higher than that of ordinary Brillouin sensors [42]. In 2012, the operation of BDG applying different spatial modes of an FMF has been demonstrated [21], where the number of possible combinations of the pump-probe pairs is increased according to the number of optical modes guided in the FMF. In this intermodal BDG operation, the grating is written by one spatial mode and used to reflect another spatial mode at an optical frequency different from the pump.

Figure 16 shows the schematics of the intermodal BDG operation in the e-core TMF, where the writing (upper) and reading (lower) procedures of the acoustic grating using the LP_{01} and LP_{11} mode are described, respectively [43]. The MSC was used to launch the pump in the LP_{01} mode for writing the BDG, and to launch and retrieve the probe in the LP_{11} mode to read out the grating reflection. The insets A and B denote the spectral relation of four optical components and the four different BDG frequencies occurring in the operation of BDG based on the e-core TMF, respectively.

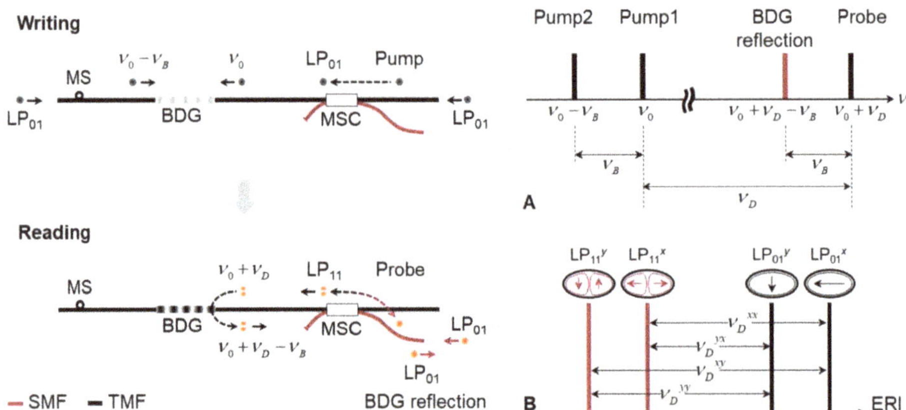

Figure 16. Schematic of the intermodal BDG operation in the e-core TMF. Inset A: Spectral relation of pump and probe waves. Inset B: Four possible pairs of pump-probe modes in the e-core TMF resulting in four different BDG frequencies [43].

The phase-matching condition of the intermodal BDG operation is that the Brillouin frequency of the pump in the LP_{01} mode is equal to that of the probe in the LP_{11} mode. One can rewrite Equation (2) for the pump and probe as follows [43]:

$$\text{Pump}: \quad \nu_B = \frac{V_a}{\Lambda} = \frac{V_a}{c}\left(n_{01}(\nu_0)\cdot\nu_0 + n_{01}(\nu_0 - \nu_B)\cdot(\nu_0 - \nu_B)\right)$$
$$\text{Probe}: \quad \nu_B = \frac{V_a}{\Lambda} = \frac{V_a}{c}\left(n_{11}(\nu_0 + \nu_D)\cdot(\nu_0 + \nu_D) + n_{11}(\nu_0 + \nu_D - \nu_B)\cdot(\nu_0 + \nu_D - \nu_B)\right) \quad (3)$$

where n_{01} (n_{11}), Λ, c, and ν_0 are the ERI of LP_{01} (LP_{11}) mode, acoustic wavelength, speed of light, and the optical frequency of pump1, respectively. After Taylor expansion and some rearrangement, one obtains a simplified expression for ν_D in this intermodal BDG as follows [43]:

$$\nu_D = \frac{n_{01}(\nu_0) - n_{11}(\nu_0)}{n_{g11}}\cdot\nu_0 \equiv \frac{\Delta n}{n_{g11}}\cdot\nu_0 \quad (4)$$

where n_{g11} is the group index of LP_{11} mode at the optical frequency of ν_0.

While the ν_D is determined only by the polarization birefringence in a PMF, the ν_D in a FMF is determined by both intermodal and polarization birefringence. The e-core TMF is also a kind of PMF due to the geometry of core, where two eigenstates of polarization exist for each spatial mode. Therefore, four different pairs of the pump-probe modes ($LP_{01}{}^x$-$LP_{11}{}^x$, $LP_{01}{}^x$-$LP_{11}{}^y$, $LP_{01}{}^y$-$LP_{11}{}^x$, and $LP_{01}{}^y$-$LP_{11}{}^y$) are possible for the intermodal BDG operation in the e-core TMF. As an example, when Δn between two spatial modes of 3.6×10^{-3} and n_{g11} of 1.45 are used at the wavelength of 1550 nm, the ν_D in the intermodal BDG is calculated to be as high as 480 GHz.

5.1. Optical Time-Domain Analysis of BDG Spectrum Based on FMF

The experimental setup for the optical time-domain analysis of intermodal BDG is depicted in Figure 17 [43], in which two DFB-LD's with center wavelengths of 1550 and 1547 nm, were used for the pump and probe wave, respectively. The output from the pump was applied to build a BOTDA configuration using an SSBM and EOM, similar to the setup presented in Figure 7, where the only difference is the change of the modulator used to sweep the frequency offset from EOM to SSBM. The duration and peak power of the pump and probe pulse were 50 and 15 ns, and 24 and 27 dBm, respectively. The power of the CW pump2 was around 15 dBm. The state of polarization of the pump1, pump2, and probe was separately controlled by a polarization controller (PC) before being launched to the FUT (i.e., e-core TMF) through an MSC. The frequency offset between the pump and probe was swept by the current control of the pump LD with a step of 4 MHz for measuring BDG spectra. The inset A shows the timing of pump and probe pulses at the end of FUT, which was controlled to maximize the signal amplitude, and the inset B is the optical spectrum measured by using an optical spectrum analyzer (OSA) at the position of the FBG where the BDG reflection is seen in the dashed box.

As described in inset B of Figure 16, the ν_D's in the intermodal BDG operation can be classified into three groups, the highest, two middles, and the lowest, corresponding to the pump-probe pairs of $LP_{01}{}^x$-$LP_{11}{}^y$, $LP_{01}{}^x$-$LP_{11}{}^x$, $LP_{01}{}^y$-$LP_{11}{}^y$, and $LP_{01}{}^y$-$LP_{11}{}^x$ modes, respectively. The polarization control of each optical wave is crucial, and one can also identify the state of polarization by monitoring the variation of reflection power according to the polarization control at specific frequency offset between the pump and probe waves. Figure 18a–d show the distribution maps of the BDG spectra for the pump-probe pairs of $LP_{01}{}^x$-$LP_{11}{}^x$, $LP_{01}{}^x$-$LP_{11}{}^y$, $LP_{01}{}^y$-$LP_{11}{}^x$, and $LP_{01}{}^y$-$LP_{11}{}^y$ modes, respectively, measured by the BDG-OTDA system [43]. The local BDG spectrum commonly has multiple peaks within a frequency span of ~2 GHz, which is thought to originate from the non-uniformity of the FUT with the spatial resolution (2 m) of the system. Gradual increase in the center frequency of local spectra along the FUT is observed in all four cases, which is thought to reflect the gradual change of the geometric parameters of e-core TMF in the preform fabrication or drawing process.

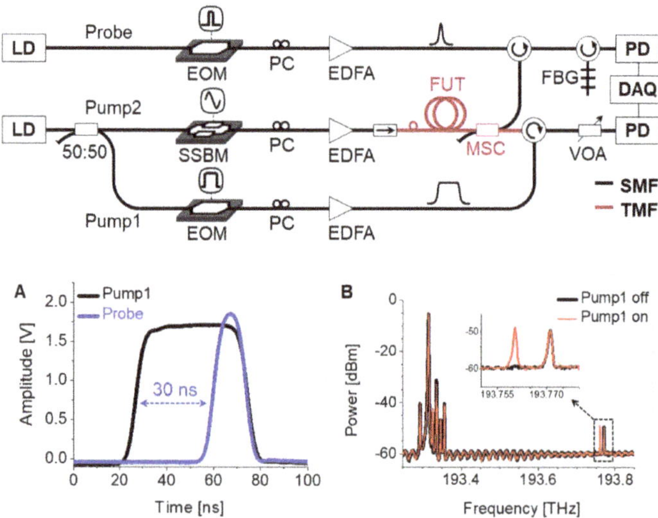

Figure 17. Experimental setup for optical time-domain analysis of intermodal BDG spectrum [43]. Inset A: Timing of the pump1 and probe pulses. Inset B: BDG spectrum measured at the position of the FBG. Reprinted with permission from Ref. [43] ©The Optical Society.

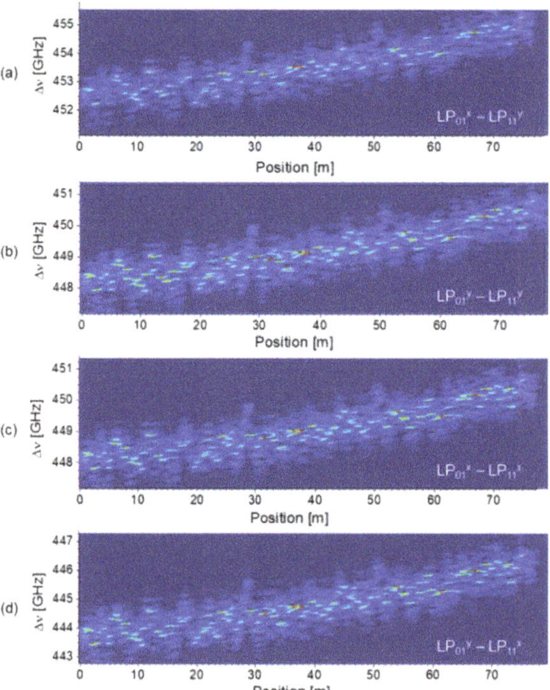

Figure 18. Distribution maps of the intermodal BDG spectra of the e-core TMF measured by the BDG-OTDA system, for the pump-probe pairs of (**a**) $LP_{01}{}^{x}$-$LP_{11}{}^{y}$, (**b**) $LP_{01}{}^{y}$-$LP_{11}{}^{y}$, (**c**) $LP_{01}{}^{x}$-$LP_{11}{}^{x}$, and (**d**) $LP_{01}{}^{y}$-$LP_{11}{}^{x}$ modes. Reprinted with permission from Ref. [43] ©The Optical Society.

The distribution maps of ν_D obtained by applying the center of mass fitting to the local BDG spectra in Figure 18 are plotted in Figure 19a, which show the variations of the intermodal birefringence along the fiber between the pump and probe modes. The distribution of polarization birefringence for each LP mode can also be acquired from the maps of ν_D by the following formulas [43]:

$$\begin{aligned}
\nu_D{}^{xy} - \nu_D{}^{xx} &= \frac{(n_{01}{}^x - n_{11}{}^y) - (n_{01}{}^x - n_{11}{}^x)}{n_{g11}} \cdot \nu_0 \equiv \frac{\Delta n_{11}}{n_{g11}} \cdot \nu_0 \\
\nu_D{}^{xy} - \nu_D{}^{yy} &= \frac{(n_{01}{}^x - n_{11}{}^y) - (n_{01}{}^y - n_{11}{}^y)}{n_{g11}} \cdot \nu_0 \equiv \frac{\Delta n_{01}}{n_{g11}} \cdot \nu_0
\end{aligned} \tag{5}$$

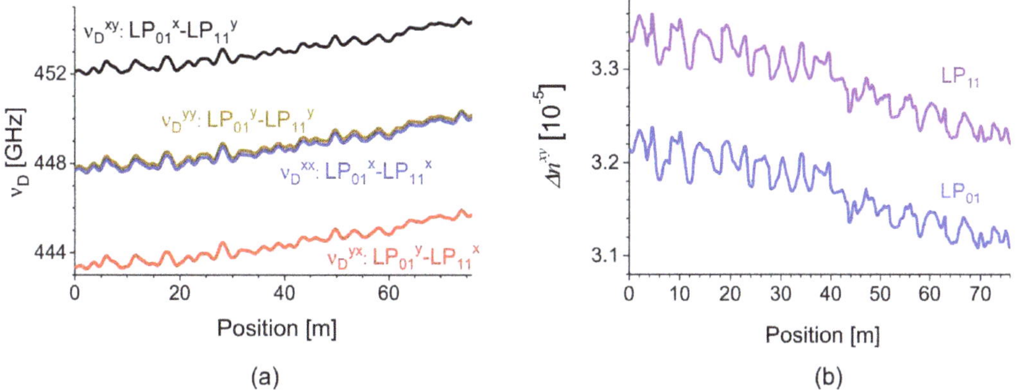

Figure 19. (**a**) Distribution map of ν_D for different pairs of pump-probe modes of intermodal BDG. (**b**) Distribution map of polarization birefringence of two LP modes reconstructed from (**a**). Reprinted with permission from Ref. [43] ©The Optical Society.

The distribution map of polarization birefringence calculated from the results of Figure 19a is plotted in Figure 19b for each LP mode, where gradual decrease of birefringence along the fiber is commonly observed as well as 1–2% local fluctuations. The polarization birefringence of LP_{11} mode is ~4% larger than that of the LP_{01} mode, while both have similar local fluctuations. Considering that the ν_D of conventional PMF's such as PANDA and bow-tie fibers, is around 45 GHz (Δn of ~3.5×10^{-4}) [45], one can see that the intermodal birefringence of the e-core TMF in Figure 19a is almost 10 times larger than the polarization birefringence of PMF's, and the contribution of the polarization birefringence in Figure 19b is less than 1% of the overall intermodal birefringence in the e-core TMF.

Figure 20a,b show the strain and temperature dependence of ν_D for different pairs of pump-probe modes, measured using two test sections near the rear end of the FUT with lengths of 2 m and 5 m for strain and temperature, respectively. The coefficients for the pump-probe pairs of $LP_{01}{}^x$-$LP_{11}{}^y$, $LP_{01}{}^y$-$LP_{11}{}^y$, $LP_{01}{}^x$-$LP_{11}{}^x$, and $LP_{01}{}^y$-$LP_{11}{}^x$ modes were -0.018, -0.012, -0.089, and -0.081 MHz/μɛ for strain and -0.16, 2.3, 2.9, and 4.9 MHz/°C for temperature, respectively. It is worth mentioning that one of the temperature coefficients is negative with the others positive, while all of the strain coefficients are negative. For references, the strain and temperature coefficients of conventional PMF's are all positive and all negative, respectively [39,40,45]. Additionally, the coefficients are overall much smaller than those of the PMF's, and the origin of which needs further investigation, although it is thought to be somewhat related to the smaller polarization birefringence of the e-core TMF.

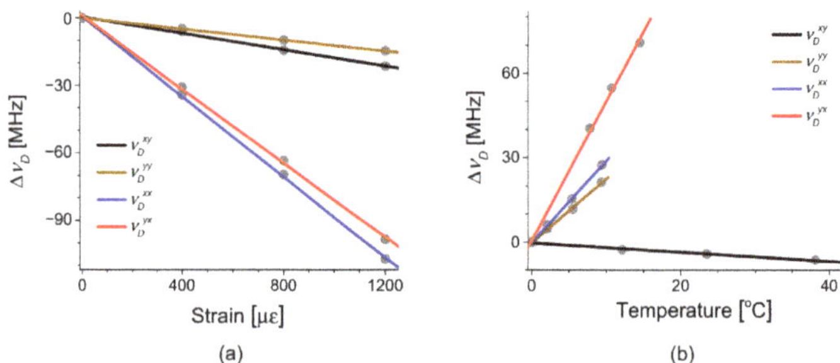

Figure 20. (**a**) Strain- and (**b**) Temperature-dependence of ν_D of intermodal BDG for different pairs of pump-probe modes in the e-core TMF. Reprinted with permission from Ref. [43] ©The Optical Society.

5.2. Optical Time-Domain Reflectometry of the BDG in a Few-Mode Fiber

The discrimination of temperature and strain variations based on BDG has been experimentally demonstrated by simultaneously measuring both ν_D and ν_B using a conventional PMF [40,41,44]. One of possible drawbacks in the distributed BDG sensor based on PMF is the requirement of two optical systems each to acquire the information on the ν_D and ν_B distribution, respectively. On the other hand, the FMF-based BDG sensor has a potential to discriminate the temperature and strain by measuring multiple ν_D's, for which only a single optical system is required, so it could provide a cost-effective solution. In addition, the analysis-type sensing systems such as BOTDA and BDG-OTDA commonly have a loop configuration, although some exceptions have been reported [46], so they require launching of optical waves to an FUT from both ends. As an alternative approach, the optical time domain reflectometry of BDG spectrum (BDG-OTDR) has been proposed in 2012, which not only allows the single end access to the FUT but also simplifies the experimental setup by removing an extra modulator [47].

Figure 21 is the schematic of the BDG-OTDR system based on the e-core TMF [48]. The BDG-OTDR does not use any microwave device since the BDG is generated by the amplified spontaneous Brillouin scattering (ASBS) of the pump pulse, while the power and duration of pump pulse should be larger to secure enough signal amplitude for detection, compared to the BDG-OTDA. Similar to the case of BDG-OTDA, the LP_{01} and LP_{11} modes were used as the pump and probe waves, each of which was selectively launched and retrieved by the MSC.

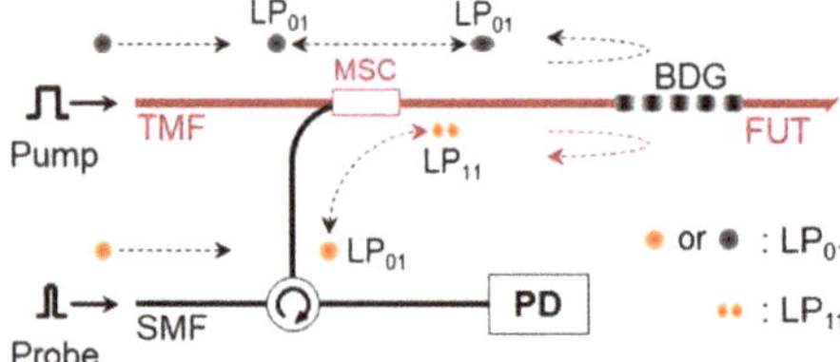

Figure 21. Schematic of the BDG-OTDR system based on the e-core TMF [48].

One can decide the best pair of pump-probe modes for intermodal BDG to perform a discriminative measurement of temperature and strain considering the condition number of transfer matrix [49], and the pump-probe pairs of $LP_{01}{}^x$-$LP_{11}{}^x$ and $LP_{01}{}^y$-$LP_{11}{}^x$ modes in the e-core TMF were applied for the demonstration of discriminative sensing in [48].

The experimental setup of BDG-OTDR is presented in Figure 22. Two different DFB-LD's, with a center wavelength of 1549 and 1546 nm, were used as light sources for the pump and probe, respectively. The pump and probe waves were modulated as a pulse with a duration of 300 ns and 20 ns, respectively, by EOM's, and the pump and probe pulses were amplified by EDFA's to the peak level of over 30 dBm. As shown in inset A, the propagations of two pulses were synchronized to maximize the signal amplitude. The optical spectrum measured at the position of FBG is shown in the inset B, where, as observed in the zoomed view, the Stokes wave of the spontaneous Brillouin scattering of probe is about 2.7 dB amplified by the intermodal BDG reflection while it is still 13 dB smaller than the Rayleigh scattering of probe. The average value of ν_D was about 448.2 GHz.

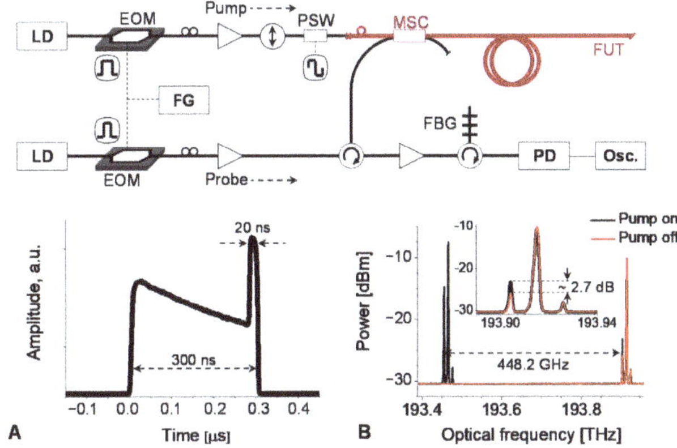

Figure 22. Experimental setup of the BDG-OTDR based on the e-core TMF. Inset A: Timing and duration of the pump (300 ns) and probe (20 ns) pulses. Inset B: Optical spectrum measured at the position of FBG. Reprinted with permission from Ref. [48] ©The Optical Society.

The local BDG spectra and their shift by the strain and temperature variations are depicted as a function of frequency offset ($\Delta\nu$) between the pump and probe in Figure 23. The BDG spectra of the $LP_{01}{}^x$-$LP_{11}{}^x$ and $LP_{01}{}^y$-$LP_{11}{}^x$ modes are shown in Figure 23a,b with temperature variations, and Figure 23c,d with strain variations, respectively [48]. It is seen that both BDG spectra move to the lower frequency when positive strain is applied in Figure 23c,d, while the spectral shift is not clear under the temperature variations in Figure 23a,b due to the arbitrary change of the relative amplitude of side peaks.

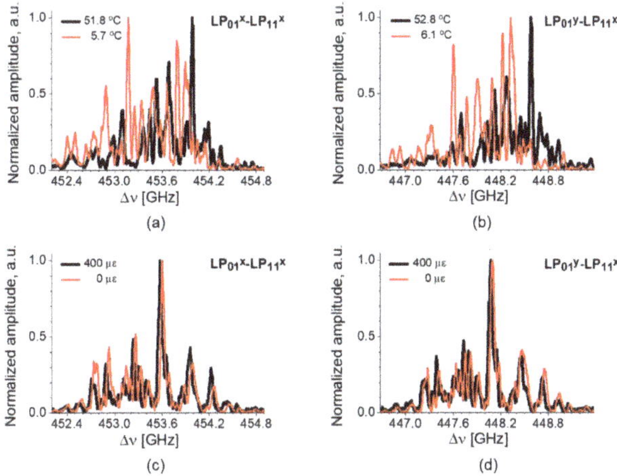

Figure 23. The shift of local BDG spectrum by temperature variations for the pump-probe pairs of (**a**) $LP_{01}{}^x$-$LP_{11}{}^x$ and (**b**) $LP_{01}{}^y$-$LP_{11}{}^x$ modes, and by strain variations for (**c**) $LP_{01}{}^x$-$LP_{11}{}^x$ and (**d**) $LP_{01}{}^y$-$LP_{11}{}^x$ modes. Reprinted with permission from Ref. [48] ©The Optical Society.

The cross-correlation fitting is an effective way to quantify the amount of shift for the multi-peak BDG spectra [50]. Figure 24a,b are examples of the cross correlation fitting applied to the experimental results presented in Figure 23a–d, respectively. The single dominant peak in Figure 24b indicates the clear shift in spectral domain under the strain variation, and it is also feasible to apply in the case of temperature variation, through the existence of the peak in Figure 24a.

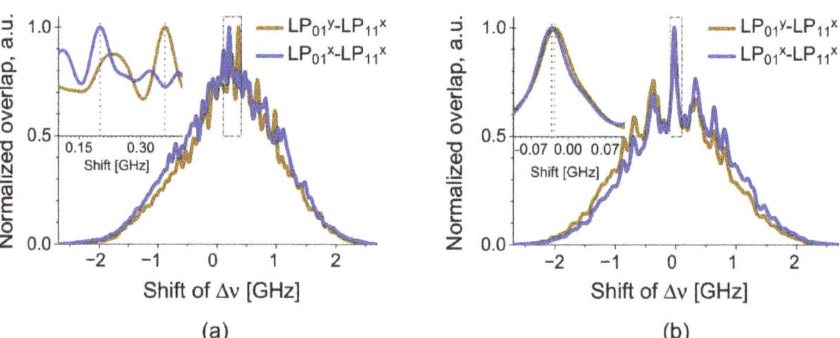

Figure 24. Examples of the cross correlation fitting applied to the multi-peak BDG spectra to quantify the change of ν_D under the variations of (**a**) temperature and (**b**) strain.

Accurate discrimination of strain and temperature variations has been a challenging task from the early stage of researches on distributed Brillouin sensors, and several techniques have been proposed applying simultaneous measurement of two independent properties [40,41,44,51]. The BDG-OTDR based on FMF also has been reported as a potential tool for the discriminative measurement, by applying temperature (C_T) and strain (C_ε) coefficients of different pairs of pump-probe modes [48]. The pump-probe pairs of $LP_{01}{}^x$-$LP_{11}{}^x$ and $LP_{01}{}^y$-$LP_{11}{}^x$ modes in the e-core TMF have shown different values of C_T, C_ε of $+ 4.3$ MHz/°C, -0.093 MHz/µε and $+ 7.6$ MHz/°C, and -0.085 MHz/µε, respectively. For the discriminative measurement of strain and temperature variations, the inverse matrix is calculated using four coefficients as follows:

$$\begin{pmatrix} \Delta\varepsilon \\ \Delta T \end{pmatrix} = \frac{1}{\det \begin{vmatrix} C_\varepsilon{}^{xx} & C_T{}^{xx} \\ C_\varepsilon{}^{yx} & C_T{}^{yx} \end{vmatrix}} \begin{pmatrix} C_T{}^{yx} & -C_T{}^{xx} \\ -C_\varepsilon{}^{yx} & C_\varepsilon{}^{xx} \end{pmatrix} \begin{pmatrix} \nu_D{}^{xx} \\ \nu_D{}^{yx} \end{pmatrix} = \begin{pmatrix} I_\varepsilon{}^{xx} & I_\varepsilon{}^{yx} \\ I_T{}^{xx} & I_T{}^{yx} \end{pmatrix} \begin{pmatrix} \nu_D{}^{xx} \\ \nu_D{}^{yx} \end{pmatrix} \quad (6)$$

The accuracy of the discriminative sensing can be evaluated by the condition number [49], and it is calculated as ~50 for the BDG-OTDR system, which is similar to that of the scheme using ν_D and ν_B with PMF [44].

Figure 25 shows the experimental results of discriminative sensing by the BDG-OTDR, where Figure 25a is the distribution map of two $\Delta\nu_D$'s under the strain and temperature variation applied at test sections near the end of a 95 m FUT [48]. The distribution map of strain and temperature variations reconstructed by the matrix calculation for each position is shown in Figure 25b, with dotted lines indicating the strain and temperature variations of 400 µε and 46.7 °C applied to the test sections. The calculated strain and temperature at those positions were 386.7 µε and 47.1 °C, respectively, and the discrepancies of 13.3 µε and 0.4 °C were less than the error range of ±105 µε and ±1.6 °C originating from the frequency drift between two light sources.

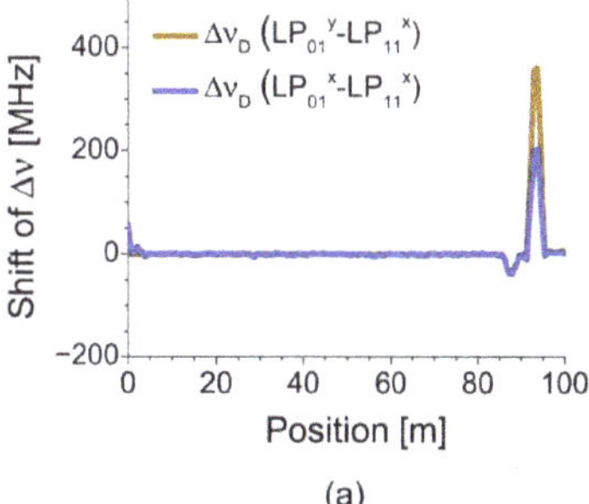
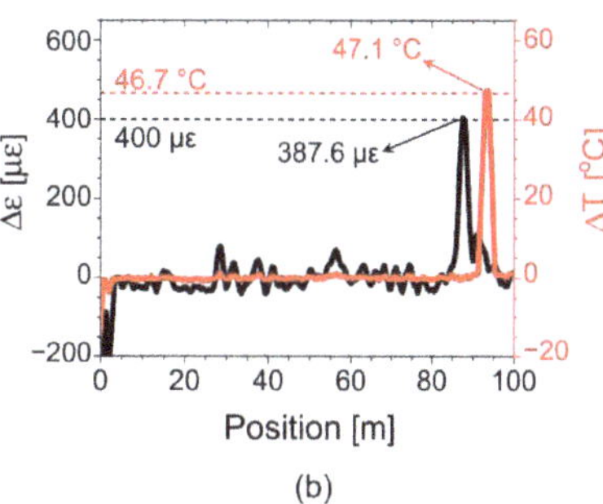

Figure 25. (**a**) Distribution map of the $\Delta\nu_D$ under temperature and strain variations applied to the test sections near the end of FUT for different pairs of pump-probe modes. (**b**) The distribution map of temperature and strain variations reconstructed from the result of (**a**) by Equation (6). Reprinted with permission from Ref. [48] ©The Optical Society.

The presented method has been a unique way so far to obtain the distribution maps of both intermodal and polarization birefringence simultaneously, which, as we believe, could be useful in designing, manufacturing, and evaluating circular-core or polarization-maintaining FMF's and related products.

6. Conclusions

Recent advances on the development of distributed Brillouin sensors based on few-mode fibers (FMF's) have been reviewed, where the experimental results on the characterization of intra- and intermodal Brillouin scatterings, Brillouin sensor systems, the characterization of intermodal Brillouin dynamic grating (BDG), and BDG-based sensor systems have been presented on the basis of the elliptical core two-mode fiber. We think the FMF is a highly flexible platform for distributed sensing, which can support various combinations of optical modes for intermodal SBS and BDG, and the use of a high-performance mode coupling device like the mode selective coupler in this work is crucial for implementing the FMF-based Brillouin or BDG sensor system.

Further research needs to be done on the design parameters of the FMF optimized for Brillouin sensors that can provide strain, temperature, and pressure coefficients sufficiently different for various intra- and intermodal SBS, and is applicable to accurate discrimination of the effects of ambient variables. The axial uniformity of the waveguide properties of e-core FMF could be a significant issue in long-range sensing applications. Also, the requirement of using specialty fiber like the e-core TMF and FMF could possibly deteriorate the practicality of FMF-based distributed sensors in terms of availability and cost. We think a commercially available FMF, such as the PANDA-type PMF at a wavelength of 2 μm, might be useful as the PANDA-FMF at the communication wavelength, providing an alternative to the e-core FMF, with possibly better axial uniformity.

Author Contributions: Y.H.K. carried out most of experiments presented in this review during his master and Ph.D studies, and K.Y.S. developed and provided the basic idea and methodology as his supervisor. All authors have read and agreed to the published version of the manuscript.

Funding: This work was supported the National Research Foundation of Korea (NSF) (20190599), and the Institute of Information & communications Technology Planning & Evaluation (IITP) grant funded by the Korea government (MSIT) (2020-0-01012).

Institutional Review Board Statement: Not applicable.

Informed Consent Statement: Not applicable.

Data Availability Statement: The data presented in this study are available on request from the corresponding author. The data are not publicly available as they involve the subsequent application of patent for invention and the publication of project deliverables.

Acknowledgments: Authors are grateful to KS Photonics for providing the mode selective couplers (MSC's) used in experiments.

Conflicts of Interest: The authors declare no conflict of interest.

References

1. Horiguchi, T.; Kurashima, T.; Tateda, M. Tensile strain dependence of Brillouin frequency shift in silica optical fibers. *IEEE Photonics Technol. Lett.* **1989**, *1*, 107–108. [CrossRef]
2. Kurashima, T.; Horiguchi, T.; Tateda, M. Distributed-temperature sensing using stimulated Brillouin scattering in optical silica fibers. *Opt. Lett.* **1990**, *15*, 1038–1040. [CrossRef] [PubMed]
3. Soto, M.A.; Bolognini, G.; Di Pasquale, F. Long-range simplex-coded BOTDA sensor over 120 km distance employing optical preamplification. *Opt. Lett.* **2011**, *36*, 232–234. [CrossRef] [PubMed]
4. Song, K.Y.; He, Z.; Hotate, K. Distributed strain measurement with millimeter-order spatial resolution based on Brillouin optical correlation domain analysis. *Opt. Lett.* **2006**, *31*, 2526–2528. [CrossRef]
5. Kim, Y.H.; Song, K.Y. Tailored pump compensation for Brillouin optical time-domain analysis with distributed Brillouin amplification. *Opt. Express* **2017**, *25*, 14098–14105. [CrossRef]
6. Wang, B.; Fan, X.; Fu, Y.; He, Z. Dynamic strain measurement with kHz-level repetition rate and centimeter-level spatial resolution based on Brillouin optical correlation domain analysis. *Opt. Express* **2018**, *26*, 6916–6928. [CrossRef]
7. Lee, C.; Chiang, P.; Chi, S. Utilization of a dispersion-shifted fiber for simultaneous measurement of distributed strain and temperature through Brillouin frequency shift. *IEEE Photonics Technol. Lett.* **2001**, *13*, 1094–1096. [CrossRef]
8. Zou, W.; He, Z.; Kishi, M.; Hotate, K. Stimulated Brillouin Scattering and Its Dependences on Temperature and Strain in a High-Delta Optical Fiber with F-Doped Depressed Inner-Cladding. *Opt. Lett.* **2007**, *32*, 600–602. [CrossRef]
9. Berdagué, S.; Facq, P. Mode division multiplexing in optical fibers. *Appl. Opt.* **1982**, *21*, 1950–1955. [CrossRef]
10. Richardson, D.J.; Fini, J.M.; Nelson, L.E. Space-division multiplexing in optical fibers. *Nat. Photon.* **2013**, *7*, 354–362. [CrossRef]
11. Song, K.Y.; Hwang, I.K.; Yun, S.H.; Kim, B.Y. High performance fused-type mode-selective coupler using elliptical core two-mode fiber at 1550 nm. *IEEE Photonics Technol. Lett.* **2002**, *14*, 501–503. [CrossRef]
12. Li, B.; Zhan, X.; Tang, M.; Gan, L.; Shen, L.; Huo, L.; Fu, S.; Tong, W.; Liu, D. Long-period fiber gratings inscribed in few-mode fibers for discriminative determination. *Opt. Express* **2019**, *27*, 26307–26316. [CrossRef]
13. Park, H.S.; Song, K.Y.; Yun, S.H.; Kim, B.Y. All-fiber wavelength-tunable acoustooptic switches based on intermodal coupling in fibers. *J. Lightw. Technol.* **2002**, *10*, 1864–1868. [CrossRef]
14. Park, K.J.; Song, K.Y.; Kim, Y.K.; Lee, J.H.; Kim, B.Y. Broadband mode division multiplexer using all-fiber mode selective couplers. *Opt. Express* **2016**, *24*, 3543–3549. [CrossRef] [PubMed]
15. Li, A.; Wang, Y.; Fang, J.; Li, M.-J.; Kim, B.Y.; Shieh, W. Few-mode fiber multi-parameter sensor with distributed temperature and strain discrimination. *Opt. Lett.* **2015**, *40*, 1488–1491. [CrossRef] [PubMed]
16. Song, K.Y.; Kim, Y.H.; Kim, B.Y. Intermodal stimulated Brillouin scattering in two-mode fibers. *Opt. Lett.* **2013**, *38*, 1805–1807. [CrossRef]
17. Song, K.Y.; Kim, Y.H. Characterization of stimulated Brillouin scattering in a few-mode fiber. *Opt. Lett.* **2013**, *38*, 4841–4844. [CrossRef]
18. Horiguchi, T.; Tateda, M. Optical-fiber-attenuation investigation using stimulated Brillouin scattering between a pulse and a continuous wave. *Opt. Lett.* **1989**, *14*, 408–410. [CrossRef] [PubMed]
19. Hotate, K.; Hasegawa, T. Measurement of Brillouin gain spectrum distribution along an optical fiber using a correlation-based technique—Proposal, experiment and simulation. *IEICE Trans. Electron.* **2000**, *E83-C*, 405–412.
20. Song, K.Y.; Zou, W.; He, Z.; Hotate, K. All-optical dynamic grating generation based on Brillouin scattering in polarization-maintaining fiber. *Opt. Lett.* **2008**, *33*, 926–928. [CrossRef]
21. Li, S.; Li, M.-J.; Vodhanel, R.S. All-optical Brillouin dynamic grating generation in few-mode optical fiber. *Opt. Lett.* **2012**, *37*, 4660–4662. [CrossRef]
22. Boyd, R.W. *Nonlinear Optics*, 3rd ed.; Academic Press: Cambridge, MA, USA, 2008.
23. Nikles, M.; Thevenaz, L.; Robert, P.A. Brillouin gain spectrum characterization in single-mode optical fibers. *J. Light. Technol.* **1997**, *15*, 1842–1851. [CrossRef]
24. Koyamada, Y.; Sato, S.; Nakamura, S.; Sotobayashi, H.; Chujo, W. Simulating and Designing Brillouin Gain Spectrum in Single-Mode Fibers. *J. Light. Technol.* **2004**, *22*, 631–639. [CrossRef]
25. Kim, B.Y.; Blake, J.N.; Huang, S.Y.; Shaw, H.J. Use of highly elliptical core fibers for two-mode fiber devices. *Opt. Lett.* **1987**, *12*, 729–731. [CrossRef]
26. Wang, Z.; Ju, J.; Jin, W. Properties of elliptical-core two-mode fiber. *Opt. Express* **2005**, *13*, 4350–4357. [CrossRef]
27. Murphy, K.; Miller, M.; Vengsarkar, A.; Claus, R. Elliptical-core two mode optical-fiber sensor implementation methods. *J. Light. Technol.* **1990**, *8*, 1688–1696. [CrossRef]
28. Yun, S.H.; Hwang, I.K.; Kim, B.Y. All-fiber tunable filter and laser based on two-mode fiber. *Opt. Lett.* **1996**, *21*, 27–29. [CrossRef]
29. Li, A.; Hu, Q.; Shieh, W. Characterization of stimulated Brillouin scattering in a circular-core two-mode fiber using optical time-domain analysis. *Opt. Express* **2013**, *21*, 31894–31906. [CrossRef]

30. Alahbabi, M.N.; Cho, Y.T.; Newson, T.P. 150-km-range distributed temperature sensor based on coherent detection of spontaneous Brillouin backscatter and in-line Raman amplification. *J. Opt. Soc. Am. B* **2005**, *22*, 1321–1324. [CrossRef]
31. Renner, H. Bending losses of coated single-mode fibers: A simple approach. *J. Light. Technol.* **1992**, *10*, 544–551. [CrossRef]
32. Kim, Y.H.; Song, K.Y. Characterization of Distributed Brillouin Sensors Based on Elliptical-Core Two-Mode Fiber. *IEEE Sens. J.* **2018**, *19*, 2155–2161. [CrossRef]
33. Nakamura, A.; Okamoto, K.; Koshikiya, Y.; Manabe, T.; Oguma, M.; Hashimoto, T.; Itoh, M. High-sensitivity detection of fiber bends: 1-mm-band mode-detection OTDR. *J. Lightw. Technol.* **2015**, *33*, 4862–4869. [CrossRef]
34. Smith, J.; Brown, A.; DeMerchant, M.; Bao, X. Pulse width dependance of the Brillouin loss spectrum. *Opt. Commun.* **1999**, *168*, 393–398. [CrossRef]
35. Jeong, J.H.; Lee, K.; Song, K.Y.; Jeong, J.-M.; Lee, S.B. Differential measurement scheme for Brillouin optical correlation domain analysis. *Opt. Express* **2012**, *20*, 27094–27101. [CrossRef] [PubMed]
36. Chin, S.; Thevenaz, L. Tunable photonic delay lines in optical fibers. *Laser Photonics Rev.* **2012**, *6*, 724–738. [CrossRef]
37. Sancho, J.; Primerov, N.; Chin, S.; Antman, Y.; Zadok, A.; Sales, S.; Thevenaz, L. Tunable and reconfigurable multi-tap microwave photonic filter based on dynamic Brillouin gratings in fibers. *Opt. Express* **2012**, *20*, 6157–6162. [CrossRef]
38. Santagiustina, M.; Chin, S.; Primerov, N.; Ursini, L.; Thévenaz, L. All-optical signal processing using dynamic Brillouin gratings. *Sci. Rep.* **2013**, *3*, 1–6. [CrossRef] [PubMed]
39. Song, K.Y.; Zou, W.; He, Z.; Hotate, K. Optical time-domain measurement of Brillouin dynamic grating spectrum in a polarization-maintaining fiber. *Opt. Lett.* **2009**, *34*, 1381–1383. [CrossRef] [PubMed]
40. Zou, W.; He, Z.; Hotate, K. Demonstration of Brillouin distributed discrimination of strain and temperature using a polarization-maintaining optical fiber. *IEEE Photonics Technol. Lett.* **2010**, *22*, 526–528. [CrossRef]
41. Dong, Y.; Chen, L.; Bao, X. High-spatial-resolution time-domain simultaneous strain and temperature sensor using Brillouin scattering and birefringence in a polarization-maintaining fiber. *IEEE Photonics Technol. Lett.* **2010**, *22*, 1364–1366. [CrossRef]
42. Kim, Y.H.; Kwon, H.; Kim, J.; Song, K.Y. Distributed measurement of hydrostatic pressure based on Brillouin dynamic grating in polarization maintaining fibers. *Opt. Express* **2016**, *24*, 21399–21406. [CrossRef]
43. Kim, Y.H.; Song, K.Y. Mapping of intermodal beat length distribution in an elliptical-core two-mode fiber based on Brillouin dynamic grating. *Opt. Express* **2014**, *22*, 17292–17302. [CrossRef] [PubMed]
44. Zou, W.; He, Z.; Hotate, K. Complete discrimination of strain and temperature using Brillouin frequency shift and birefringence in a polarization-maintaining fiber. *Opt. Express* **2009**, *17*, 1248–1255. [CrossRef] [PubMed]
45. Kim, Y.H.; Song, K.Y. Characterization of nonlinear temperature dependence of Brillouin dynamic grating spectra in polarization-maintaining fibers. *J. Lightw. Technol.* **2015**, *33*, 4922–4927. [CrossRef]
46. Nikles, M.; Thevenaz, L.; Robert, P.A. Simple distributed fiber sensor based on Brillouin gain spectrum analysis. *Opt. Lett.* **1996**, *21*, 758–760. [CrossRef]
47. Song, K.Y. High-sensitivity optical time-domain reflectometry based on Brillouin dynamic gratings in polarization maintaining fibers. *Opt. Express* **2012**, *20*, 27377–27383. [CrossRef] [PubMed]
48. Kim, Y.H.; Song, K.Y. Optical time-domain reflectometry based on a Brillouin dynamic grating in an elliptical-core two-mode fiber. *Opt. Lett.* **2017**, *42*, 3036–3039. [CrossRef]
49. Jin, W.; Michie, W.C.; Thursby, G.; Konstantaki, M.; Culshaw, B. Simultaneous measurement of strain and temperature: Error analysis. *Opt. Eng.* **1997**, *36*, 598–610. [CrossRef]
50. Farahani, M.A.; Castillo-Guerra, E.; Colpitts, B.G. Accurate estimation of Brillouin frequency shift in Brillouin optical time domain analysis sensors using cross correlation. *Opt. Lett.* **2011**, *36*, 4275–4277. [CrossRef]
51. Song, K.Y.; Kim, J.H.; Yoon, H.J.B. Simultaneous measurement of temperature and strain distribution by optical time-domain analysis of Brillouin dynamic grating. In Proceedings of the 15th OptoElectronics and Communications Conference (OECC2010), Sapporo, Japan, 5–9 July 2010.

Article

Towards Detecting Red Palm Weevil Using Machine Learning and Fiber Optic Distributed Acoustic Sensing

Biwei Wang [1,2,†], **Yuan Mao** [1,†], **Islam Ashry** [1,†], **Yousef Al-Fehaid** [3], **Abdulmoneim Al-Shawaf** [3],
Tien Khee Ng [1], **Changyuan Yu** [2] and **Boon S. Ooi** [1,*]

[1] Computer, Electrical and Mathematical Sciences and Engineering (CEMSE) Division, King Abdullah
     University of Science and Technology (KAUST), Thuwal 23955-6900, Saudi Arabia;
     biwei.wang@kaust.edu.sa (B.W.); yuan.mao@kaust.edu.sa (Y.M.); islam.ashry@kaust.edu.sa (I.A.);
     tienkhee.ng@kaust.edu.sa (T.K.N.)
[2] Department of Electronic and Information Engineering, The Hong Kong Polytechnic University,
     Hong Kong, China; changyuan.yu@polyu.edu.hk
[3] Center of Date Palms and Dates, Ministry of Environment, Water and Agriculture,
     Al-Hassa 31982, Saudi Arabia; alfehaid2000@gmail.com (Y.A.-F.); yassir1418@yahoo.com (A.A.-S.)
*   Correspondence: boon.ooi@kaust.edu.sa
†   These authors contributed equally to this work.

Abstract: Red palm weevil (RPW) is a detrimental pest, which has wiped out many palm tree farms worldwide. Early detection of RPW is challenging, especially in large-scale farms. Here, we introduce the combination of machine learning and fiber optic distributed acoustic sensing (DAS) techniques as a solution for the early detection of RPW in vast farms. Within the laboratory environment, we reconstructed the conditions of a farm that includes an infested tree with ~12 day old weevil larvae and another healthy tree. Meanwhile, some noise sources are introduced, including wind and bird sounds around the trees. After training with the experimental time- and frequency-domain data provided by the fiber optic DAS system, a fully-connected artificial neural network (ANN) and a convolutional neural network (CNN) can efficiently recognize the healthy and infested trees with high classification accuracy values (99.9% by ANN with temporal data and 99.7% by CNN with spectral data, in reasonable noise conditions). This work paves the way for deploying the high efficiency and cost-effective fiber optic DAS to monitor RPW in open-air and large-scale farms containing thousands of trees.

Keywords: red palm weevil; fiber optic acoustic sensing; machine learning

Citation: Wang, B.; Mao, Y.; Ashry, I.; Al-Fehaid, Y.; Al-Shawaf, A.; Ng, T.K.; Yu, C.; Ooi, B.S. Towards Detecting Red Palm Weevil Using Machine Learning and Fiber Optic Distributed Acoustic Sensing. *Sensors* **2021**, *21*, 1592. https://doi.org/10.3390/s21051592

Academic Editor: Min Yong Jeon

Received: 20 January 2021
Accepted: 20 February 2021
Published: 25 February 2021

Publisher's Note: MDPI stays neutral with regard to jurisdictional claims in published maps and institutional affiliations.

1. Introduction

The date palm is a high-value fruit crop that provides healthy nutrition security to millions of people around the world [1]. It is further considered an important source of export revenue for rural smallholders worldwide. Unfortunately, the date production and trade are in danger because of the red palm weevil (RPW), also named as Rhynchophorus ferrugineus [2,3]. RPW is a Coleopteran snout pest, which is considered the single most destructive pest of palm trees. Young and soft trees aged less than 20 years, which represent ~50% of the total cultivated date palm trees, are vulnerable since RPW typically targets them [4]. Beside date palms, RPW also attacks coconut, oil, and ornamental palms [4,5]. In the past few decades, RPW has been found in more than 60 countries including the Mediterranean region, parts of Central America, Middle East, North Africa, among others [4,6]. This plague has globally destroyed many palm farms causing severe economic losses in a form of lost production or pest-control costs. As representative examples, Figure 1 shows the RPW's impact on two date palm trees, after treating the trees with scraping to entirely remove the RPW.

Figure 1. Two representative examples of treated trees by scraping to remove red palm weevil (RPW).

In the early stage of infestation, palm trees can be healed with chemical treatments [7]. However, a palm tree only shows visual distress signs in a well-advanced stage of infestation, where it is difficult to save the tree. Many techniques have been reported in the literature for the early detection of RPW [8–10]. Some detection methods such as x-ray based tomography [9] and trained dogs [10] are accurate; however, they lack the feasibility in large-scale farms because of their slow scanning processes. The most promising early detection methods are based on sensing the larvae sound, while they are chewing the core of a palm trunk. The larvae start to produce eating sound in an early infestation stage, where the larvae are less than two weeks old [11]. Existing acoustic detection technologies rely on inserting acoustic probes within the individual tree trunks and building a wireless network to communicate with the sensors [8]. Unfortunately, it is cost-ineffective to assign a sensor per tree, especially for vast farms containing thousands of trees. Additionally, this method is invasive and may harm the trees or create nests for insects.

We recently reported a solution of using a fiber optic distributed acoustic sensor (DAS) for the early detection of RPW, such that a single optical fiber is noninvasively wound around the palm trees to possibly scan a large-scale farm within a short time [11,12]. As reported in [11], distinguishing the healthy and infested trees was achieved through a straightforward signal processing algorithm since the experiment was carried out in a controlled environment. Identifying infested trees in open-air farms, where the optical fiber might be subjected to harsh environmental noises, would require a more advanced signal processing technique to classify the larvae sound and other noise sources.

To pave the way for utilizing the fiber optic DAS to monitor real farms, here, we introduce neural network-based machine learning algorithms to classify healthy and infested trees, based on the data collected by a fiber optic DAS. Within a laboratory environment, we mimic the environmental conditions of a farm that includes the healthy/infested palm tree and other noise sources. In particular, a sound of ~12 day old weevil larvae is played inside a tree trunk and meanwhile, the tree is subjected to external wind and bird sounds as noise sources. A fully-connected artificial neural network (ANN) [13] and a convolutional neural network (CNN) [14] are used to recognize the infested and healthy trees in the noisy environment. We further investigate the impact of different optical fiber's jackets on mitigating the external noise around the trees. This work would be highly beneficial towards the future deployment of the fiber optic DAS for the early detection of RPW in vast real farms.

2. Experimental Design

Figure 2a shows the overall design of our experiment. The optical and electronic components of the DAS system are assembled within a sensing unit, whose contents will be described shortly later. The output light from the sensing unit is launched into a single-mode fiber (SMF) and we wind a section of the fiber around a tree trunk. Within the trunk, we implant a loudspeaker (SRS-XB21, Sony, 20 Hz–20,000 Hz frequency transmission range) that continuously plays an eating sound of ~12 day old weevil larvae. At a ~1 m distance from the tree, there is a fan that blows air, with a speed of ~3 m/s, towards the optical fiber and tree. Additionally, also at a ~1 m distance from the tree, we locate another loudspeaker

(AudioCube, Allococac, 40 Hz–20,000 Hz frequency transmission range) that continuously produces bird sounds. At the outer surface of the tree, we place a sound level meter that can record the various sound intensity levels used in the experiment. The measured sound intensity level of the background noise within the laboratory is ∼51 dB, caused by the instruments working in the laboratory, which raises to ∼71 dB when only playing the bird sounds. The intensity level of the bird sounds is roughly equal to that we hear in farms. According to the literature [8] and also our experience, humans can hear the larvae sound under acceptable environmental noise. As measured by the meter, when only the loudspeaker within the tree is turned on, we set the intensity levels of the larvae sound to be within the range [∼51 dB–∼75 dB]. The low-level cannot be distinguished from the background noise by the meter, while the other high-threshold is obvious. The selected larvae sound's intensity levels within this range should represent all possible degrees of infestation (weak, medium, and strong infestation).

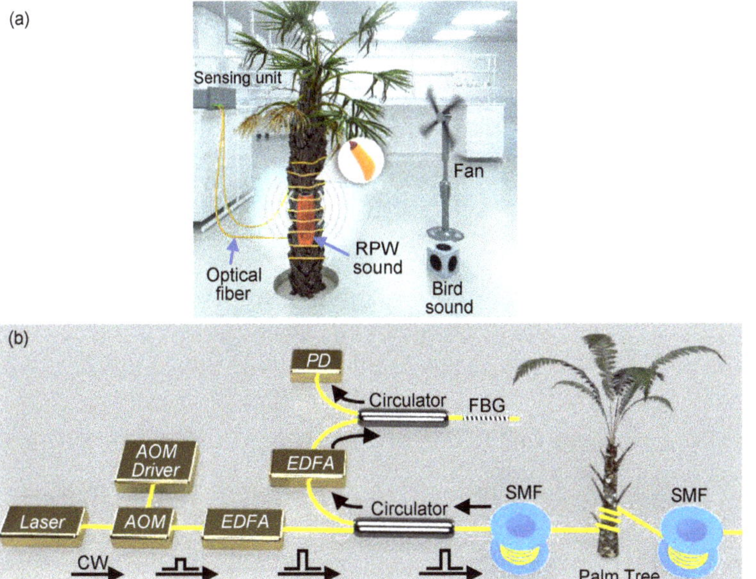

Figure 2. (**a**) Overall design of using the fiber optic distributed acoustic sensor (DAS) to detect the RPW sound. (**b**) Experimental setup of the DAS interrogation unit.

The sensing unit comprises the interrogation system of the fiber optic DAS (Figure 2b), which is designed using the phase-sensitive optical time-domain reflectometry (ϕ-OTDR) [15]. A narrow linewidth laser generates a continuous wave (CW) light of a 100 Hz linewidth and a 40 mW optical power. The laser light is then converted into optical pulses using an acousto-optic modulator (AOM) that produces pulses of a 50 ns width and a 5 kHz repetition rate. The selected pulse width offers a DAS system of a 5 m spatial resolution. The power of the optical pulses is amplified using an erbium-doped fiber amplifier (EDFA), while its output light is launched through a circulator into a standard SMF of a ∼2 km length. At a ∼1 km distance from the input port of the SMF, we wind a 5 m section of the fiber around the tree trunk. The backscattered signal from the SMF is amplified with another EDFA, which amplified spontaneous emission (ASE) noise is discarded using a fiber Bragg grating (FBG). The filtered Rayleigh signal is detected by a photodetector (PD) and sampled by a digitizer of a 200-MHz sampling rate. Finally, the Rayleigh signals are recorded as 1-s periods (5000 traces per period). This experiment includes utilization of two separate standard SMFs, protected with different jackets of a 900 μm diameter (Thorlabs,

SMF-28-J9, denoted as "JKT1") and a 5 mm diameter (YOFC, YOFC-SCTX3Y-2B1-5.0-BL, denoted as "JKT2"), respectively.

Figure 3 shows an example of a Rayleigh trace recorded by the fiber optic DAS system. The high-power signal located at the start of the SMF is typical and it corresponds to the Fresnel reflection from the front facet of the SMF. In the ideal scenario, when there is no refractive index perturbation along the SMF, the shape of the Rayleigh trace remains stationary in the time-domain for all the spatial points along the entire fiber [15,16]. Consequently, the differences between the temporal subsequent Rayleigh traces and an initial reference one would be ideally zeros. In contrast, the presence of a larvae sound within the tree trunk can modulate the fiber's refractive index at the tree position, which results in changing the corresponding temporal Rayleigh signal only at the tree location. By applying the normalized differential method [17] and fast Fourier transform (FFT) to the temporal Rayleigh traces, the location of an infested tree and the larvae sound's frequencies can be identified, respectively.

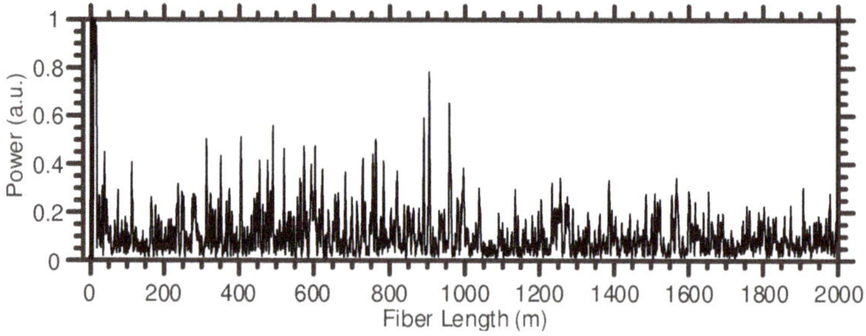

Figure 3. A representative example of a Rayleigh trace recorded by the fiber optic DAS.

3. Investigating the Impact of the Noise Sources on the DAS System

In this section, we explore the possible ways of mitigating the environmental noises, such as wind and bird sounds, which may degrade the performance of the fiber optic DAS system when detecting the RPW. The suggested techniques of reducing the noise include applying a spectral band-pass filter to alleviate the noise level within the recorded signals and further trying various optical fiber's jackets which might be shaken because of the wind. This investigation is necessary not only to improve the performance of the machine learning algorithms during classifying the healthy and infested trees but also to make the DAS system more feasible for real RPW detection.

Firstly, the spectral components of the actual larvae sound are explored. In particular, a commercial voice recorder (ICD-UX570, Sony, 50 Hz–20,000 Hz frequency response) is implanted inside a truly infested tree trunk and next to ~12 day old larvae, shown in Figure 4a, such that the voice recorder stores the larva's sound using the uncompressed linear pulse-code modulation (LPCM) format to always have a pristine quality audio file [18]. We select this specific RPW life stage to examine if our sensor can detect the larvae sound at an early stage, so that the palm tree can still be saved and cured. During the recording time, the larvae are eating and moving naturally within the trunk without any restriction. Consequently, the quality of the simulated sound in the laboratory should be comparable to the real one. The age of the larvae can be well controlled via an artificial infestation process [11], where it is carried out in a secured research facility to avoid spreading the RPW to other healthy trees. Interestingly, it is observed that the larvae almost continuously produce the sound while they are chewing the tree trunk. Figure 4b shows two different representative examples of the larvae sound's power spectra, where each corresponds to a 0.5 s recording interval. The results of Figure 4b indicate that the majority of the larvae sound's optical power is around 400 Hz.

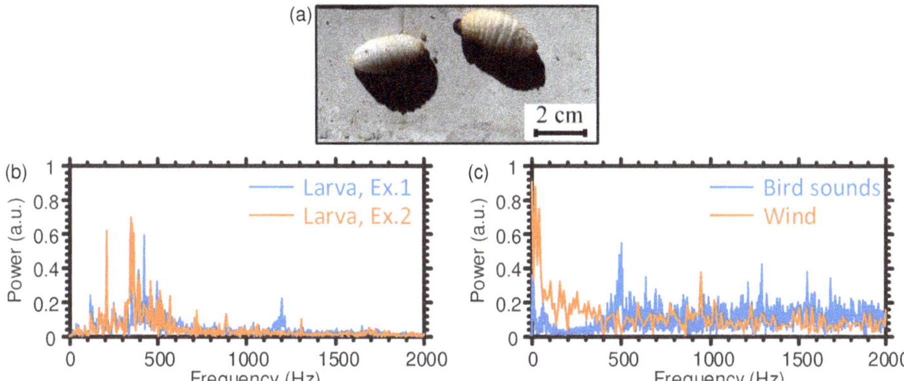

Figure 4. (**a**) ~12 days old weevil larvae. (**b**) Two representative examples of the larvae sound's spectra, marked as Ex. 1 and Ex. 2, produced using the data of the voice recorder. (**c**) Examples of the power spectra produced by bird sounds and wind.

In contrast, the used bird sounds have a broad spectrum (Figure 4c, blue line), which interferes with that of the larvae. Regarding the wind as a noise source, the orange line in Figure 4c represents an example of the vibration's power spectrum caused by wind when shaking the "JKT2" fiber. The vibration caused by wind is dominated by the tree swinging, which has low-frequency components. However, wind may also directly shake the optical fiber to produce other vibrations of high frequencies, which may overlap with that of the larvae sound. This is because the vibration strength caused by the wind is larger than that of the larvae sound; therefore, the wind's high-frequency vibration has to be taken into account as a noise source. Given the results of Figure 4b,c, for the entire following temporal vibration data that we collect using the fiber optic DAS, we will apply a [200 Hz–800 Hz] band-pass filter to enhance the signal-to-noise ratio (SNR) of our system. This is because the spectral filter can discard the low vibration frequency components, less than 200 Hz, to cancel the inevitable mechanical vibration in the laboratory and the tree swinging caused by wind. Meanwhile, it filters out the high-frequency (larger than 800 Hz) components, produced by the electronic/optical components in the system, without impacting the larvae sound's dominant frequencies (around 400 Hz).

Focusing on the experimental design of Figure 2a, we initially switch off the fan and the outside noise loudspeaker, while we only play the larvae sound using the loudspeaker implanted inside the tree trunk. Figure 5a–d show two representative examples of the normalized differential time-domain signals recorded using the DAS system [17], followed by applying the [200 Hz–800 Hz] band-pass filter, when using the SMF of JKT1 [JKT2]. Clearly, the two fibers accurately locate the position of the infested tree at the ~1 km distance from the input ports of the fibers. The other noisy signals, which sometimes appear at the start of the SMFs, are a result of the fiber front facet's reflection.

Next, the two loudspeakers are switched off and we only turn on the fan to inspect the impact of the wind on the two SMFs. Wind would be considered the primary noise source in open-air farms, especially because our detection technique is noninvasive and the fiber would typically be subjected to vibrations caused by wind. Even with applying the [200 Hz–800 Hz] band-pass filter, the SMF of JKT1 is impacted by the wind to produce temporal vibrations as those shown in Figure 6a,b. The low frequency vibrations, produced by tree swinging as a result of the wind, can be easily discarded with filtering out the frequencies below 200 Hz [19]. However, blowing the wind to directly hit the optical fiber results in shaking the fiber with frequencies that rely on the thickness and material of the fiber's jacket. As shown in Figure 6a,b, the SMF of JKT1 that has a relatively small diameter (900 μm) produces vibration signals, caused by the wind, which may resemble those of the larvae sound. This behavior may confuse the machine learning algorithms during distinguishing the healthy and infested trees.

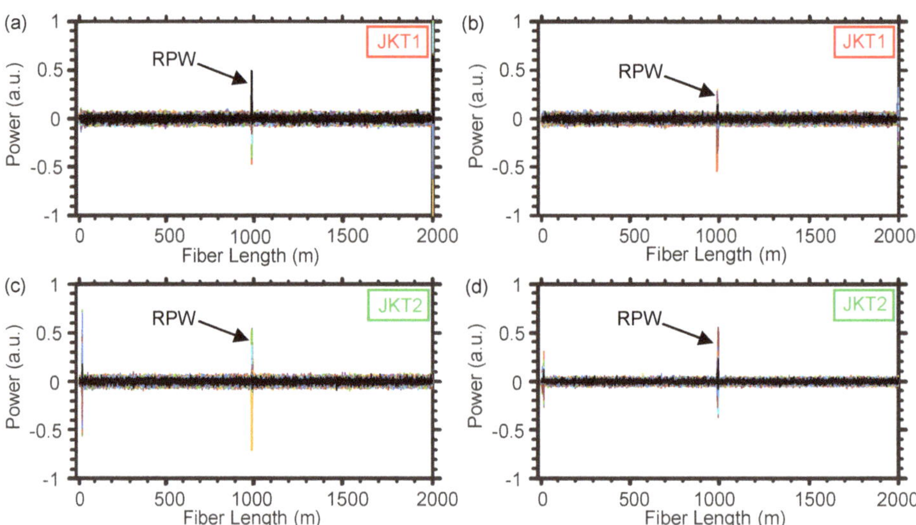

Figure 5. Representative examples of the infested tree's position information when using the SMF of JKT1 (**a**), (**b**) and JKT2 (**c**), (**d**).

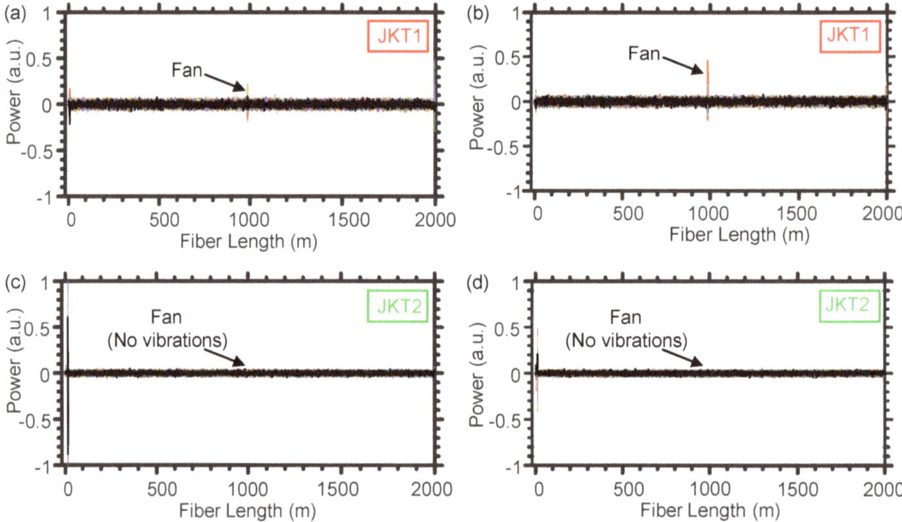

Figure 6. Representative examples of the temporal vibrations caused by the wind when using the SMF of JKT1 (**a**), (**b**) and JKT2 (**c**), (**d**).

Similarly, while switching off the two loudspeakers and turning only the fan on, we use the DAS system when winding the SMF of JKT2 around the tree trunk. Since the JKT2 is relatively thick (5 mm diameter), the fiber rarely generates shaking frequencies within the [200 Hz–800 Hz] range because of the wind [Figure 6c,d]. Such comparison between the two fiber jackets in terms of mitigating the noise produced by wind is crucial for determining the proper optical fiber cable that can be used in the future in real farms. Besides, compared with JKT1, JKT2 has an additional advantage that it is durable enough to sustain the harsh environmental conditions of farms and the SMF inside JKT2 cannot be easily broken by, for example, stepping on the fiber by farmers.

We further investigate the impact of the noise produced by bird that may surround the optical fiber in farms. In particular, we switch off the larvae sound's loudspeaker and the fan, while playing only the outside loudspeaker. Fortunately, the two SMFs of JKT1 and JKT2 cannot "hear" the bird sounds, as shown respectively in Figure 7a,b. This is because the air between the loudspeaker and the optical fiber jackets significantly attenuates the vibration energy of the bird sounds. Typically, a fiber optic DAS system can be used to sense sound propagating through the air by utilizing a thin metallic sheet, attached to the fiber, to amplify that attenuated vibration energy by air [20]. In our experiment, however, we do not use a metallic sheet to avoid recording the acoustic noise signals generated around the tree.

It is also worth discussing the impact of using the JKT1 and JKT2 on the overall noise floor. The noise floor depends on many factors such as temporal pulse intensity fluctuation, laser phase noise and frequency drift, low extinction ratio of the launched pulses, photodetector thermal, and shot noise [15,21], which are all common when using the JKT1 or JKT2. However, another major factor that contributes to the noise floor is the overall isolation of the optical fiber from externally induced vibrations. Consequently, the thicker jacket (JKT2) typically provides a lower noise floor.

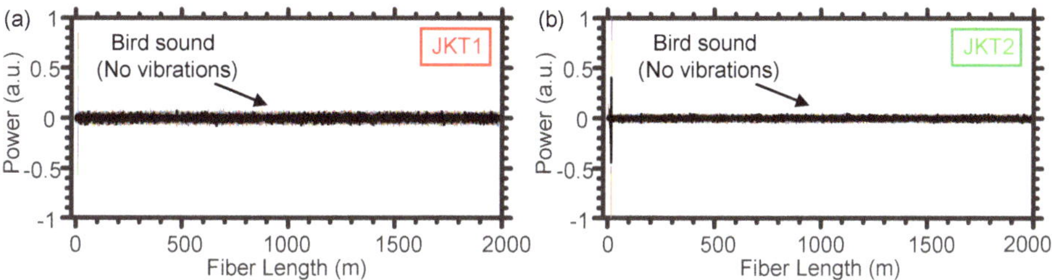

Figure 7. Representative examples of the temporal vibrations caused by the bird sounds when using the SMF of JKT1 (**a**) and JKT2 (**b**).

4. Classifying Infested and Healthy Trees Using Machine Learning Methods

Machine learning methods trained through supervised learning can be effective approaches for identifying the infested and healthy trees. Machine learning can reveal patterns associated with the larvae sound and simultaneously deal with the large amount of data produced by the DAS system. In this work, we compare the efficiencies of classifying the healthy and infested trees when using the time- and frequency-domain data as separate inputs to neural networks, which are designed using the fully-connected ANN and CNN architectures. Given the aforementioned advantages of the SMF of JKT2, we decide to use it in the subsequent analyses of classifying the healthy and infested trees using machine learning methods.

We initially focus on the way of organizing and labeling the time- and frequency-domain data for the ANN. As aforementioned, we wind a 5 m section of the fiber around the tree, while the digitizer is sampling the data at a 200 MHz frequency. Consequently, given the time-of-flight within the OTDR sensing system, the optical fiber section around the tree is represented by 10 spatial points, i.e., the digitizer's sampling resolution is ~0.5 m. For each point, the digitizer reading takes a 1 s period, i.e., 5000 readings in the time-domain per one reading period because the pulse repetition rate is 5 kHz. Since the digital band-pass filter typically distorts a short-interval at the beginning of the time-domain signal, we discard the first 250 time-domain readings for each spatial point. Thus, the collected temporal data in each trial are organized as a vector of 47,500 length (concatenating 4750 time-domain readings $\times 10$ spatial points). In contrast, by applying the FFT to the time-domain data of each spatial point, we get 2375 frequency components. Subsequently, we organize the spectral data of each trial as a vector of 23,750 length (concatenating 2375 frequency components $\times 10$ spatial points).

We label the data as "infested" or "healthy" tree, based on the SNR value of the acoustic signal at the tree position. We define the SNR as the ratio between the root-mean-square (RMS) value of the time-domain signal at the tree position and that at another calm reference fiber section of a 5 m length. We evaluate the ability of the machine learning algorithms to classify the infested and healthy trees in two cases, without and with the presence of wind. Considering the first case when ignoring the wind, we play the loudspeaker within the tree trunk and stop the fan to mark the signals of the infested tree. If the SNR > 2 dB, the minimum acceptable SNR of a DAS system [17], we record and label the signal as "infested". We collect 2000 examples of the infested signals, when the sound of the larvae loudspeaker is set at various intensity levels within the aforementioned [$\sim$51 dB–$\sim$75 dB] range. In contrast, we record other 2000 samples for the "healthy" signals, when the larvae loudspeaker and fan are off. The "healthy" signal examples are recorded regardless of the SNR value to be higher or lower than the 2 dB threshold.

Focusing on labeling the data when considering the presence of the wind, we turn on simultaneously the larvae loudspeaker and the fan to record the examples of the "infested" signals. Similarly, we record 2000 various examples when the SNR values exceed the 2 dB threshold. Next, we switch off the larvae speaker while keeping the fan on in order to record the other 2000 examples, regardless of the SNR values, for the healthy tree.

The ANN models used to handle the time- and frequency-domain data have a similar architecture, which is shown in Figure 8. This structure consists of one input layer, two hidden layers, and one output layer. The number of nodes in the input layer matches the number of elements in the data vectors, i.e., 47,500 and 23,750 for the time- and frequency-domain data, respectively. Besides, the first and second hidden layers respectively comprise 500 and 50 nodes, determined by repeated and sufficient trials to maximize the classification accuracy. At the end of the fully-connected ANN, there is an output layer of one node for the binary classification (infested or healthy). Regarding the activation functions, we use the rectified linear unit (ReLU) for the hidden layers and the sigmoid function for the output layer.

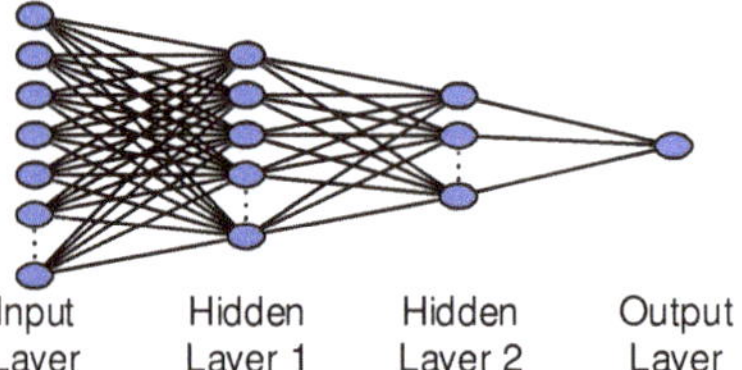

Figure 8. The ANN structure for detecting the RPW infestation using the temporal/spectral data.

When the wind is ignored (the fan is turned off), we split the collected temporal/spectral data as 60% (2400 examples) training, 20% (800 examples) validation, and 20% (800 examples) testing datasets. In this scenario, Figure 9a,c shows the evolution of the training/validation accuracy and loss with the epoch, when using the temporal [spectral] data. At the end of the training cycles, validation accuracy values of 82.0% and 99.8% are produced for the time- and frequency-domain data, accordingly. When using the temporal data [Figure 9a], the final validation accuracy is obviously lower than that of the training process, which indicates the model cannot be generalized. In contrast, as shown in Figure 9c of the spectral data, the validation accuracy perfectly matches with the training one, which confirms that the ANN model learns the features well, instead of just remembering the input data.

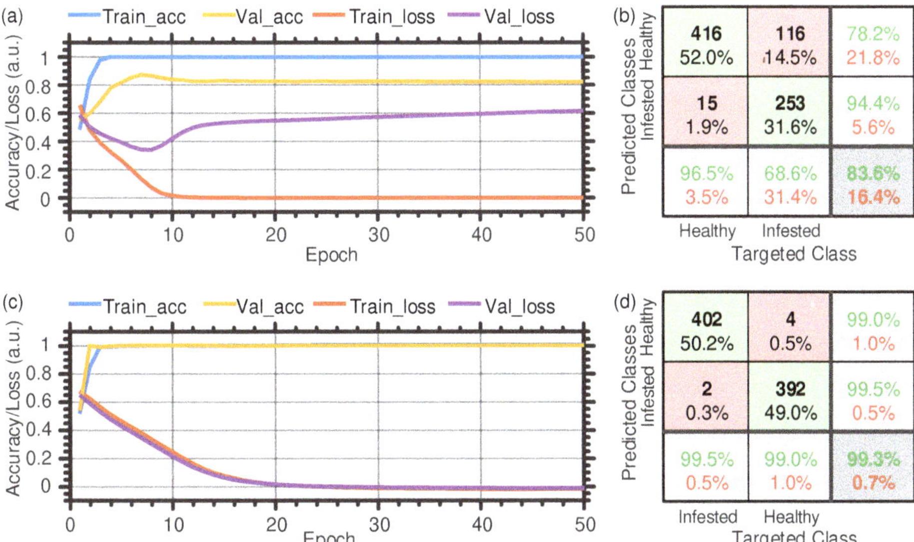

Figure 9. Training and validation history (**a**)/(**c**) and confusion matrix (**b**)/(**d**) when ignoring the wind and using the temporal/spectral dataset with the ANN. Train_acc: training accuracy; Val_acc: validation accuracy; Train_loss: training loss; Val_loss: validation loss.

Following the training and validation processes, we use the testing datasets to estimate the performance of the two models. Figures 9b,d show the confusion matrices when using the time- and frequency-domain data, respectively. In general, a confusion matrix comprises four main indices denoted as true negatives (TN), false negatives (FN), false positives (FP), and true positives (TP), which compare the actual target values with those predicted by the machine learning model [22]. Besides, some other important performance metrics (accuracy, precision, recall, and false alarms) are also included in the confusion matrix and defined as [22]:

$$
\begin{aligned}
Accuracy &= (TP + TN)/(TP + FP + TN + FN), \\
Precision &= TP/(TP + FP), \\
Recall &= TP/(TP + FN), \\
False\,Alarm &= FP/(TP + FP).
\end{aligned}
$$

As shown in the confusion matrices of Figure 9b,d, the temporal data provides a total classification accuracy of 83.6%, while that of the spectral data is 99.3%. The entire ANN's performance parameters are summarized in Table 1, first and second rows, when neglecting the wind and using the temporal and spectral data. As a result, to get a high distinguishing accuracy between the infested and healthy trees, it is recommended to use the ANN with the spectral data of the larvae sound. This is attributed to that the chewing sound of the larvae can be shifted within the 1 s recording frame, which makes it difficult for the ANN model to learn using a limited dataset. However, the shifted temporal acoustic signals produce similar spectra, which facilitate the classification process using the frequency-domain data. Given this conclusion, we decide to rely on the spectral components with the ANN to analyze the subsequent more complex scenario, when the wind impact is considered.

We split again the spectral data, collected when the fan is turned on, as 60% (2400 examples) training, 20% (800 examples) validation, and 20% (800 examples) testing datasets. After the training and validation processes, the spectral testing dataset is used to examine the performance of the trained model. For this scenario, the third row of Table 1 shows a summary of the ANN's performance results. The ANN model provides a total classification accuracy of 99.6%, which is slightly higher than that produced in the case of neglecting the wind. The precision, recall, and false alarm rates also show minor improvements. These results indicate that the ANN model can perfectly learn the larvae sound's spectral pattern in the two scenarios, without and with wind, while the tiny perturbations caused by the wind slightly increases the robustness and generalization ability of the model.

Table 1. Summary of the ANN's performance for the various environmental conditions.

Data	Accuracy	Precision	Recall	False Alarm
Temporal data, without wind	83.6%	94.4%	68.6%	5.6%
Spectral data, without wind	99.3%	99.5%	99.0%	0.5%
Spectral data, with wind	99.6%	99.7%	99.5%	0.3%
Spectral data, combined	99.9%	99.9%	99.9%	0.1%

A more realistic case to consider is combining the two spectral datasets, with and without the wind as a noise source. This is reasonable since the air blows intermittently in real farms. Thus, we merge the two datasets to have in total 8000 examples for the infested and healthy trees. Again, we split the entire data as 60% (4800 examples) training, 20% (1600 examples) validation, and 20% (1600 examples) testing datasets. When using the combined data, the classification accuracy, precision, recall, and false alarm rates are improved [fourth row of Table 1], as compared with those of the two former separate cases. These results indicate that the performance of the ANN model is enhanced given the large quantity and more variety of the training data. Thus, one can conclude that the ANN model performs excellently when using the combined spectral data of the more realistic scenario in farms; however, the ANN model has a relatively poor performance with the temporal input data.

CNNs are popular deep neural network structures, designed to be spatially invariant [14]. In other words, they are not sensitive to the position of the features, which would be effective in handling the temporal larvae sound that is shifting in the time-domain. In addition, compared with the fully-connected ANNs, CNNs have relatively less parameters to train, which makes CNNs easier and more efficient to train with the same quantity of datasets [14,23]. Since CNNs have proven high efficiency in classifying images, we arrange the temporal and spectral data in two-dimensional matrix forms. In particular, the time- and frequency-domain examples are arranged as 10 (spatial points) × 4750 (temporal readings) and 10 (spatial points) × 2375 (spectral components), respectively. As representative examples, Figure 10a–d show the CNN model's input images of the (a) temporal and "infested", (b) temporal and "healthy", (c) spectral and "infested", and (d) spectral and "healthy" data, respectively.

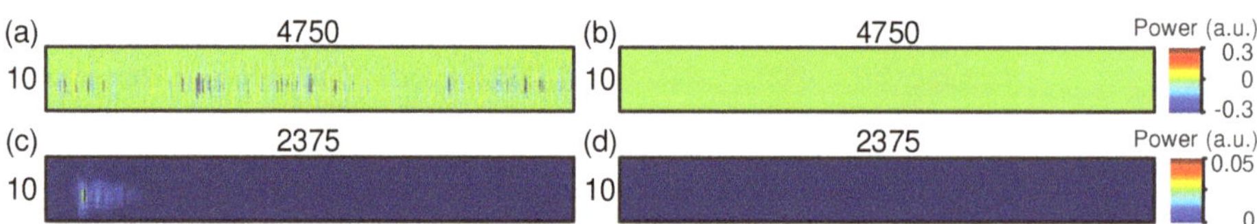

Figure 10. CNN model's input images of the (**a**) temporal and "infested", (**b**) temporal and "healthy", (**c**) spectral and "infested", and (**d**) spectral and "healthy" data.

Figure 11 shows the architecture of the CNN model used to separately handle the temporal and spectral input data. The architecture respectively comprises an input layer, two pairs of convolutional and max pooling layers, a flatten layer, a fully-connected layer, and an output layer. The first (second) convolutional layer has the ReLU activation function and includes 32 (64) filters of a 3×50 (3×5) size and 1×25 (1×5) stride. The two max pooling layers have the same 2×2 pool size and 2×2 stride. After the flatten layer, the fully-connected layer has the ReLU activation function and consists of 50 nodes. Similar to the ANN, the output layer of the CNN also has one node with sigmoid activation function for the purpose of binary classification (healthy or infested).

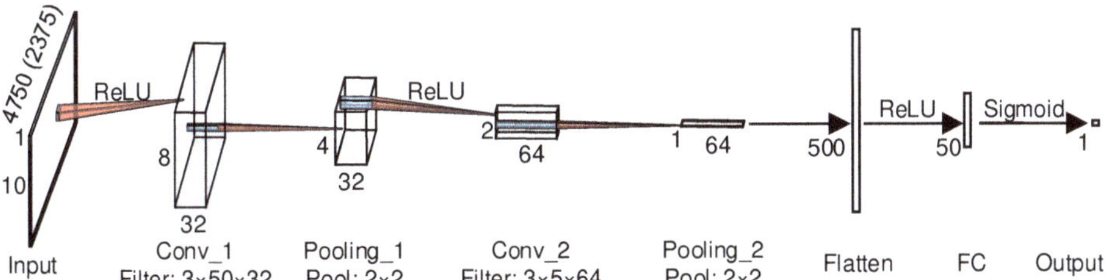

Figure 11. The CNN structure for detecting the RPW infestation using the temporal/spectral data. Conv: convolutional; FC: fully-connected.

In terms of the data labeling and splitting for the CNN model, we adopt the same techniques and data quantity as those used with the fully-connected ANN. Considering the ideal scenario when the wind is ignored (the fan is turned off), Figure 12a,c show the evolution of the training/validation accuracy and loss with the epoch for the temporal and spectral data, respectively. After finishing the training cycles, validation accuracy values of 100% and 99.5% are accordingly obtained when using the time- and frequency-domain data. Besides, the two confusion matrices when using the temporal and spectral testing datasets are shown in Figure 12b,d, respectively. The results of the confusion matrices show that the performance of the CNN with the temporal data is excellent with 100.0% accuracy, while that of the spectral data is slightly lower (99.3%). Clearly, as compared with the results of Figure 9, the CNN significantly improves the classification efficiency in the time-domain. This proves the aforementioned two main advantages of the CNN model over the fully-connected ANN model, i.e., the CNN's spatial invariance and less parameters to train. These results are important since using CNN would offer a real-time detection of RPW, without the need to apply the intensive FFT to the time-domain data.

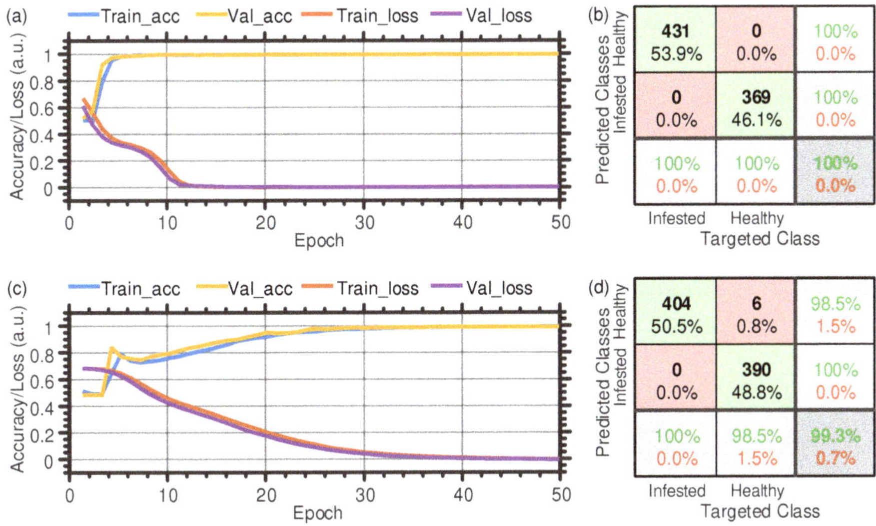

Figure 12. Training and validation history (**a**)/(**c**) and confusion matrix (**b**)/(**d**) when ignoring the wind and using the temporal/spectral dataset with the CNN. Train_acc: training accuracy; Val_acc: validation accuracy; Train_loss: training loss; Val_loss: validation loss.

Table 2 summarizes the CNN's performance when using the temporal and spectral data, in case of ignoring or considering the wind, or mixing the two scenarios. As can be observed, the CNN model has a superior performance in the various situations with a minimum classification accuracy of 98.3%. Taking into consideration that the time-domain data is easier to process, compared with the spectral data that require additional FFT step, we recommend using the CNN and time-domain data for the feasible detection of RPW. Given this conclusion and considering the more reasonable case of the combined data, the CNN with the temporal data provides 99.7% accuracy, 99.5% precision, 99.9% recall, and 0.5% false alarm [third row, Table 2]. The high precision and low false alarm values confirm the reliability of the CNN model in classifying the healthy and infested trees. On the other side, the high recall value represents the great ability and sensitivity of the CNN model to locate the "infested" signals from a mixed "healthy" and "infested" set.

Table 2. Summary of the CNN's performance for the various environmental conditions.

Data	Accuracy	Precision	Recall	False Alarm
Temporal data, without wind	100.0%	100.0%	100.0%	0.0%
Temporal data, with wind	99.9%	99.7%	100.0%	0.3%
Temporal data, combined	99.7%	99.5%	99.9%	0.5%
Spectral data, without wind	99.3%	100.0%	98.5%	0.0%
Spectral data, with wind	98.3%	99.5%	97.0%	0.5%
Spectral data, combined	99.1%	99.7%	98.3%	0.3%

5. Discussion

We have examined the possibility of using machine learning and fiber optic DAS to distinguish healthy and infested trees, in the laboratory environment. In farms, however, many financial and technical issues have to be considered during the practical implementation. As reported in the literature [24], the sensing range of the fiber optic DAS can be typically extended to ~10 km with a spatial resolution down to 1 m. Assuming the separation between two consecutive trees is ~10 m and we wind a ~1 m fiber section around each tree, a DAS sensing unit can simultaneously monitor ~1000 trees. The entire cost of our DAS system including the optical fiber cable is ~US$37,000; thus, the monitoring cost per tree is ~US$37. A recommended deployment plan comprises permanent installation of the optical fiber cables in farms, because fiber optics are relatively cheap and easy to be plugged in/out of the DAS unit, while sharing a portable DAS sensing unit between the farms. Such a plan can significantly reduce the monitoring cost per tree, as the 5 m fiber section, wound around a tree in our experiment, costs only ~US$2. Alternatively, if the farms are close to each other, time-division-multiplexing (TDM) [25] could be adopted by connecting the farms' optical fibers with a single sensing unit via an optical switch, to scan the individual farms in different time frames.

Besides, in a real palm tree farm, the optical fiber cable might be broken because of the farming activities around the trees. Fortunately, the SMF (YOFC, YOFC-SCTX3Y-2B1-5.0-BL) is well protected with metallic rods to make the fiber shockproof. Furthermore, the outer jacket layer of the fiber optic cable can sustain temperatures as high as 60 °C, which helps the fiber to "survive" in the farms' harsh environments. For more protection, we recommend burying the optical fiber in the soil between the trees. In the worst-case scenario, if the fiber is getting broken for any reason, the OTDR system can accurately locate the fault point with the spatial resolution of the system [26]. A portable fusion splicer can be easily used to fix the optical fiber on-site.

In case of adopting the spiral winding of the optical fiber around trees in a farm, an annual maintenance to the sensing system would be needed because of the trees' girth growing. Planning ahead for having redundant fiber lengths between trees will help readjust the fiber wraps around the trees later as they grow. Alternative to the spiral winding plan, longitudinal zigzag attachment of the fiber to the tree trunk is a backup strategy; however, a heavy-duty and stretchable plastic wrap would be used as an outer layer to provide sufficient contact between the optical fiber and tree trunk. The latter strategy is part of our plans for future work on this topic.

It is also worth discussing the conditions of contact between the fiber optic cable and the tree trunk. The spatial resolution of our fiber optic DAS system is 5 m, which indicates the system cannot distinguish the spatial separations between vibration events that occur within the 5 m distance. In case of winding a fiber length shorter than 5 m around the tree, vibration events occurring close to the tree may not be distinguished from those at the tree, resulting in generating false alarms. Thus, the minimum length of fiber needed in this experiment is 5 m. Besides, in our experiment, we wind a 5 m fiber optic section around the tree with moderate tightness. We observed that some points along the fiber section are not directly touching the tree trunk, because of the sharpness of the trunk at some locations; however, the DAS system still efficiently detects the larvae sound. In other words, having all the points along the 5 m fiber optic section in direct and firm contact with the trunk is not necessary to make the system works. In general, the more tightly we wind the fiber optic around the tree, the higher the SNR. However, it is experimentally difficult to quantify the relationship between the SNR and the tightness of winding the optical fiber. Additionally, winding a fiber section of length longer than 5 m would improve the performance of the DAS system. However, the 5 m spatial resolution we use already provides excellent classification accuracy.

This discussion shows that the main advantage of our sensor, compared with those reported in the literature, is that the fiber optic DAS provides distributed detection of RPW. Within a short time, an entire farm would be scanned, which saves time and effort as compared with the other detection methods [8–10] that inspect the trees individually. However, a drawback of our reported system is that the initial installation of the optical fiber requires time and effort, especially in vast farms. Fortunately, the fiber installation only needs to be carried out once per farm, and then the fiber can remain in the farm permanently.

6. Conclusions

Fully-connected ANN and CNN are used to classify an infested tree with RPW and a healthy tree, using the data provided by a fiber optic DAS. To mimic the farm environment within the laboratory, the weevil larvae sound is played inside a tree trunk, while wind and bird sounds are used as noise sources around the tree. Considering the common scenario when the wind blows discretely, the ANN performs perfectly with the frequency-domain data to offer a 99.9% classification accuracy. In contrast, for the same conditions of wind, the CNN produces 99.7% and 99.1% classification accuracy when using the temporal and spectral data, respectively. Although the CNN's performance is excellent with both kinds of data, we recommend using the CNN with the temporal data to avoid the intensive FFT calculations, required to get the spectral components. The results of this work are significantly beneficial for the next step of deploying the fiber optic DAS for the early detection of RPW in real farms.

Author Contributions: Conceptualization, I.A. and Y.M.; methodology, I.A., Y.M., B.W., Y.A.-F., and A.A.-S.; software, I.A., Y.M., and B.W.; validation, I.A., Y.M., and B.W.; formal analysis, I.A., Y.M., and B.W.; investigation, I.A., Y.M., and B.W.; resources, Y.A.-F. and A.A.-S.; data curation, I.A. and B.W.; writing—original draft preparation, I.A.; writing—review and editing, T.K.N., Y.M., and B.W.; visualization, Y.M. and B.W.; supervision, B.S.O., C.Y., and T.K.N.; project administration, B.S.O.; funding acquisition, B.S.O. All authors have read and agreed to the published version of the manuscript.

Funding: This research was funded by KAUST-Research Translation Funding (REI/1/4247-01-01).

Institutional Review Board Statement: Not applicable.

Informed Consent Statement: Not applicable.

Data Availability Statement: The data presented in this study are available on reasonable request from the corresponding author. The data are not publicly available due to privacy.

Conflicts of Interest: The authors declare no conflict of interest.

References

1. Al-Shahib, W.; Marshall, R.J. The fruit of the date palm: its possible use as the best food for the future? *Int. J. Food Sci. Nutr.* **2003**, *54*, 247–259.
2. Al-Dosary, N.M.; Al-Dobai, S.; Faleiro, J.R. Review on the management of red palm weevil Rhynchophorus ferrugineus Olivier in date palm Phoenix dactylifera L. *Emir. J. Food Agric.* **2016**, *28*, 34–44.
3. El-Mergawy, R.; Al-Ajlan, A. Red palm weevil, Rhynchophorus ferrugineus (Olivier): economic importance, biology, biogeography and integrated pest management. *J. Agric. Sci. Technol. A* **2011**, *1*, 1–23.
4. Food Chain Crisis. Available online: http://www.fao.org/food-chain-crisis/how-we-work/plant-protection/red-palm-weevil/en/ (accessed on 14 Decmber 2020).
5. Wahizatul, A.A.; Zazali, C.; Abdul, R.; Nurul'Izzah, A.G. A new invasive coconut pest in Malaysia: The red palm weevil (Curculionidae: Rhynchophorus ferrugineus). *Planter* **2013**, *89*, 97–110.
6. Ferry, M.; Gomez, S. The red palm weevil in the Mediterranean area. *Palms* **2002**, *46*, 172–178.
7. Llácer, E.; Miret, J.A.J. Efficacy of phosphine as a fumigant against Rhynchophorus ferrugineus (Coleoptera: Curculionidae) in palms. *Span. J. Agric. Res.* **2010**, *3*, 775–779.
8. Rach, M.M.; Gomis, H.M.; Granado, O.L.; Malumbres, M.P.; Campoy, A.M.; Martín, J.J.S. On the design of a bioacoustic sensor for the early detection of the red palm weevil. *Sensors* **2013**, *13*, 1706–1729.
9. Haff, R.; Slaughter, D. Real-time X-ray inspection of wheat for infestation by the granary weevil, Sitophilus granarius (L.). *Trans. ASAE* **2004**, *47*, 531.
10. Suma, P.; La Pergola, A.; Longo, S.; Soroker, V. The use of sniffing dogs for the detection of Rhynchophorus ferrugineus. *Phytoparasitica* **2014**, *42*, 269–274.
11. Ashry, I.; Mao, Y.; Al-Fehaid, Y.; Al-Shawaf, A.; Al-Bagshi, M.; Al-Brahim, S.; Ng, T.K.; Ooi, B.S. Early detection of red palm weevil using distributed optical sensor. *Sci. Rep.* **2020**, *10*, 3155.
12. Ooi, B.S.; Ashry, I.; Mao, Y. Beetle Detection Using Optical Fiber Distributed Acoustic Sensor. US Patent Application No. 17/050,116, 9 July 2018.
13. Meireles, M.R.; Almeida, P.E.; Simões, M.G. A comprehensive review for industrial applicability of artificial neural networks. *IEEE Trans. Ind. Electron.* **2003**, *50*, 585–601.
14. Schmidhuber, J. Deep learning in neural networks: An overview. *Neural Networks* **2015**, *61*, 85–117.
15. Bao, X.; Zhou, D.P.; Baker, C.; Chen, L. Recent development in the distributed fiber optic acoustic and ultrasonic detection. *J. Light. Technol.* **2016**, *35*, 3256–3267.
16. Lu, Y.; Zhu, T.; Chen, L.; Bao, X. Distributed vibration sensor based on coherent detection of phase-OTDR. *J. Light. Technol.* **2010**, *28*, 3243–3249.
17. Ashry, I.; Mao, Y.; Alias, M.S.; Ng, T.K.; Hveding, F.; Arsalan, M.; Ooi, B.S. Normalized differential method for improving the signal-to-noise ratio of a distributed acoustic sensor. *Appl. Opt.* **2019**, *58*, 4933–4938.
18. Lipshitz, S.P.; Vanderkooy, J. Pulse-Code Modulation—An Overview. *J. Audio Eng. Soc.* **2004**, *52*, 200–215.
19. Mao, Y.; Ashry, I.; Ng, T.K.; Ooi, B.S. Towards Early Detection of Red Palm Weevil using Optical Fiber Distributed Acoustic Sensor. In *Optical Fiber Communication Conference*; Optical Society of America: San Diego, CA, USA, 2019; , pp. W2A.15.
20. Mao, Y.; Ashry, I.; Alias, M.S.; Ng, T.K.; Hveding, F.; Arsalan, M.; Ooi, B.S. Investigating the performance of a few-mode fiber for distributed acoustic sensing. *IEEE Photonics J.* **2019**, *11*, 1–10.
21. Bao, X.; Chen, L. Recent progress in distributed fiber optic sensors. *sensors* **2012**, *12*, 8601–8639.
22. Wu, H.; Chen, J.; Liu, X.; Xiao, Y.; Wang, M.; Zheng, Y.; Rao, Y. One-Dimensional CNN-Based Intelligent Recognition of Vibrations in Pipeline Monitoring With DAS. *J. Light. Technol.* **2019**, *37*, 4359–4366.
23. O'Shea, K.; Nash, R. An Introduction to Convolutional Neural Networks. *arXiv* **2015**, arXiv:1511.08458.
24. Chen, D.; Liu, Q.; He, Z. High-fidelity distributed fiber-optic acoustic sensor with fading noise suppressed and sub-meter spatial resolution. *Opt. Express* **2018**, *26*, 16138–16146.
25. Spirit, D.M.; Ellis, A.D.; Barnsley, P.E. Optical time division multiplexing: Systems and networks. *IEEE Commun. Mag.* **1994**, *32*, 56–62.
26. Tateda, M.; Horiguchi, T. Advances in optical time domain reflectometry. *J. Light. Technol.* **1989**, *7*, 1217–1224.

Article

The Sensitivity Improvement Characterization of Distributed Strain Sensors Due to Weak Fiber Bragg Gratings

Konstantin V. Stepanov [1,*], Andrey A. Zhirnov [1,2], Anton O. Chernutsky [1],
Kirill I. Koshelev [1], Alexey B. Pnev [1], Alexey I. Lopunov [2] and Oleg V. Butov [2]

[1] Bauman Moscow State Technical University, 2-nd Baumanskaya 5-1, 105005 Moscow, Russia;
     a.zh@bmstu.ru (A.A.Z.); chernutsky.a@bmstu.ru (A.O.C.); koshelev@bmstu.ru (K.I.K.);
     pniov@bmstu.ru (A.B.P.)
[2] Kotelnikov Institute of Radioengineering and Electronics of RAS, Mokhovaya 11-7, 125009 Moscow, Russia;
     aley@mail.ru (A.I.L.); obutov@mail.ru (O.V.B.)
*   Correspondence: stkv@bmstu.ru

Received: 26 October 2020; Accepted: 9 November 2020; Published: 11 November 2020

Abstract: Weak fiber Bragg gratings (WFBGs) in a phase-sensitive optical time-domain reflectometer (phi-OTDR) sensor offer opportunities to significantly improve the signal-to-noise ratio (SNR) and sensitivity of the device. Here, we demonstrate the process of the signal and noise components' formation in the device reflectograms for a Rayleigh scattering phi-OTDR and a WFBG-based OTDR. We theoretically calculated the increase in SNR when using the same optical and electrical components under the same external impacts for both setups. The obtained values are confirmed on experimental installations, demonstrating an improvement in the SNR by about 19 dB at frequencies of 20, 100, and 400 Hz. In this way, the minimum recorded impact (at the threshold SNR = 10) can be reduced from 100 nm per 20 m of fiber to less than 5 nm per 20 m of fiber sensor.

Keywords: weak fiber Bragg gratings; distributed fiber optic sensor; phi-OTDR

---

1. Introduction

Distributed fiber-optic monitoring systems based on phase-sensitive optical time-domain reflectometers (phi-OTDR) are gaining popularity. Their indisputable advantages include the ability to control objects of considerable length (pipelines, perimeters) automatically, allowing real-time detection of the magnitude and location of impacts on the sensor. They operate by analyzing backscattered radiation from a pulse of a highly coherent radiation source. In this case, the sensor has a high sensitivity in analyzing the phase deviations of the received signal. The received signal—the reflectogram—remains unchanged until the moment of any impact on the sensor. For example, the system can receive information about vibration effects, the analysis of which can be used to register a person's passages, excavations, vehicle movements, and other activities. The history of such systems began in the 1990s with the works of the Taylor group [1–5]. Gradually, such systems have become more common given the development of fiber optics. Canadian scientists have paid much attention to polarization effects and signal processing using wavelets [6–8]. Scientists from Spain, led by H.F. Martins, worked on various aspects of the system's operation, including implementing them, increasing the sensor length using Raman amplifiers, and measuring temperature using chirped pulses [9–12]. In Russia, several scientific groups are also engaged in this issue, with particular attention to the device's principles and limitations, the possibilities of using it as a temperature sensor, and the development of principles for restoring the phase of the recorded signal [13–20]. Many results have found practical applications in commercial systems, including in the oil and gas industry [21,22]. In the last 10 years, many scientific papers have been published that comprehensively considered methods of increasing the range of systems operation, signal processing, various applications, and tracking the frequency offset of a laser source [23–26]. Some of these works were devoted to numerically measuring the strain exerted on the fiber. However, the physical nature of the signal formation by a phase-sensitive reflectometer based on Rayleigh scattering makes signal changes nonlinear and nonsinusoidal, including with monotonic external influences. A significant step toward explicitly representing the relationship between external influences and signal changes would be to modify sensors with WFBGs. Patents for such a sensor device first appeared a long time ago [27], but they have only received practical implementation relatively recently [28–30]. Before this point, gratings at different wavelengths were mainly surveyed, or wavelength shifts were monitored [31,32]. In such a case, only characteristics (temperature or strain) of the WFBG contact area can be measured. A system with 843 FBGs in one fiber was experimentally demonstrated [32]. Since 2015, articles about WFBGs with phi-OTDR interrogation have become regular. Scientific groups in China carried out the first works surveying WFBGs, similarly to phase-sensitive reflectometry. A group from Shandong and Wuhan Universities experimentally showed a work of setup with 500 WFBGs. They achieved the pressure detection limit of 0.122 Pa in a frequency range from 450 to 600 Hz [33]. Scientists from Nanjing University in a few publications showed 11 dB enhancement in SNR with the help of WFBGs and described a quantitative strain measurement with 6.2 nε maximum error [33–37]. They also theoretically showed thelimitations of WFBG's quantity in one sensor due to attenuation on an array of them [35,36]. Since then, many studies have been carried out. Works have been presented on the possibility of restoring the impact phase by using Mach–Zehnder [38] and Michelson [39] interferometers in the recording branch. It can achieve a fiber length variation sensitivity of 117 pm/$\sqrt{\text{Hz}}$ with a distance between WFBGs equal to 99 m [38]. Investigations of vast arrays of WFBG were carried out in a laboratory: setup with 964 WFBGs for nε-scale deformation measurements [40], 90 WFBGs helped to detect 14.63 nm deformations of 50 m fiber between neighboring reflectors [41]. Parallel measurements using traditional sensors were carried out: phi-OTDR system with 660 WFBGs was compared to geophones and showed good agreement in signal shape [42]. Simultaneously, the measurement results in such systems have made it possible to obtain deformation values, including in microstrains—that is, showing the elongation of the fiber [40]. RMS measurement error of fewer than 1.32 nε was demonstrated in a frequency range from 8 to 1 kHz [43]. This is in agreement with the results of the other group: static strain resolution of 1.89 nε and dynamic resolution of 97.5 pε/$\sqrt{\text{Hz}}$ over 1 Hz [44]. Possibilities have been investigated for increasing the sensor interrogation rate three times using measurements at several wavelengths generated by AOMs with different modulation frequencies [45]. Additionally, the first studies were carried out to compare the recorded signals of a fixed frequency and the amplitude on systems with WFBGs and with a conventional phase-sensitive reflectometer in the spectral region. Exceeding of the peak level over the noise level was improved by 27 dB in an experiment with a 1 kHz impact produced by PZT [46]. In 2019, the practical application of such a system was shown for the first time. Phi-OTDR with WFBGs was used for subway monitoring. This can detect intrusion events, e.g., illegal excavator work [47]. The first steps were done to the next generation sensors, based on scattering dots. These can be fs-inscribed into fiber, which is technological in comparison to WFBG producing technology. The first experiments were conducted to show the possibility of signal improvement, similar to WFBG [48]. It was also demonstrated that a combination of frequency and phase domain time-gated coherent reflectometry methods based on a sensor with artificial reflectors can measure the distance between them [49]. For spacing between dots in fiber 9.84 ± 0.04 m, the measurement error was 0.16 μm. But this technology still needs improvement due to the instability of dot's reflection coefficients produced in one fiber.

This study was a theoretical analysis and experimental comparison of the growth in signal-to-noise ratio at the same fiber deformations in systems based on a classical phase-sensitive reflectometer scheme versus a scheme using WFBG. For this purpose, it is necessary to disassemble the signal-formation process and determine the primary sources of noise. The relevance of this problem is due to the peculiarity of signal registration by a real system in the field, which depends on a large number of parameters, including the type of fiber, the cable design, and the characteristics of the soil in which the cable is laid. On average, for a classical scheme with a phase-sensitive reflectometer, the passage and the performance of any manual work (such excavation) is believed to be detectable at a distance of no more than 5 m from the sensor cable [50–52]. At the same time, due to technological features, laying the sensor cable within just such a distance from the monitoring object is often difficult or impossible, which results in the distance from the sensor to the potential impact site exceeding the indicated 5 m, so that the event is not registered. To solve this problem, the minimum detectable level of cable deformation must be lowered. This can be done, for example, by adding hydrogel to the cable for better transmission of vibration effects directly to the fiber, thus increasing the sensor's sensitivity. Nevertheless, even if the sensitivity of the cable itself is improved, some sources of system noise cannot be eliminated. In a phase-sensitive reflectometer, the primary sources of noise are:

1. Frequency instability of the laser source, which affects the stability of the interference backscattered signal's formation;
2. Signal-spontaneous noise from the preamplifier (beats of the signal with spontaneous emission of the preamplifier within the optical filter's transmission), which is several orders of magnitude higher than the preamplifier's spontaneous-spontaneous noise (the beats of the spectral components of the preamplifier's spontaneous emission within the optical filter's transmission) due to the use of narrow-band filters in the receiving line [53];
3. The noise of the photodetector module.

In this paper, we consider the influence of low-reflective fiber Bragg gratings on these components of the parasitic noise. Thus, for example, the presence of a WFBG can remove one main source of noise: the erbium preamplifier.

2. Theory

2.1. Signal Formation in a Phase-Sensitive Reflectometer

A phase-sensitive reflectometer is a device operating based on the effects of Rayleigh scattering. The scheme is shown in Figure 1. The radiation from a highly coherent source (a laser) is brought to the required power in an amplifier (EDFA). Next, probing pulses are formed in the optical modulator (OM), which are directed through the circulator in the forward path to the sensor fiber, with the scattered radiation from the reverse path is directed to the preamplifier (pEDFA) to increase the power. The preamplifier's spontaneous emission is suppressed with an optical filter (F). Next, the radiation is registered by a photodetector (PD), digitized on an analog–digital converter (ADC), and sent for processing on a computer (PC). Information about an impact on the fiber is extracted from the instability of the reflectograms, specifically the intensity distribution of the backscattered radiation along the sensor cable.

If the radiation from the source of the probing signals has a coherence length much longer than the pulse duration, then the backscattered waves will not be added as the integrated powers, but as the amplitudes which take the phases into account. In this case, the scattering centers are any inhomogeneities in the fiber core, which are located in the glass structure chaotically and with high density. As a result, the distribution of the backscattered waves' amplitudes obeys Gauss's law, and the distribution of phases is uniform over the interval from 0 to 2π, as shown in Figure 2.

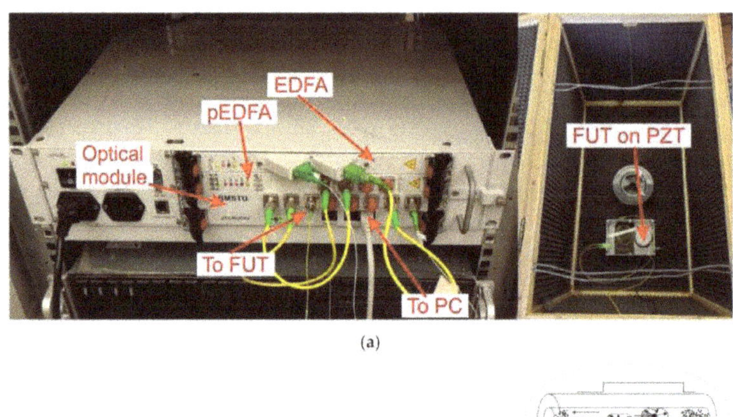

(**a**)

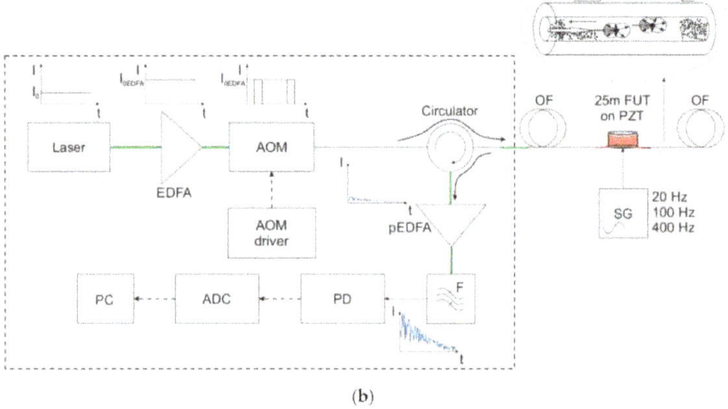

(**b**)

Figure 1. Photo and scheme of an experimental phase-sensitive reflectometer for determining a sensor system's sensitivity based on Rayleigh backscattering. Laser: radiation source, EDFA: optical erbium amplifier, OM: optical modulator, OF: optical fiber, FUT: fiber section for testing impacts, pEDFA: optical erbium preamplifier, F: optical filter, PD: photodiode, ADC: analog–digital converter, PC: personal computer. Thin solid lines—fiber optic connections inside the optical module. Bold solid lines—fiber connections outside the case. Dashed lines—electrical connections. (**a**): photo of setup; (**b**): scheme of setup.

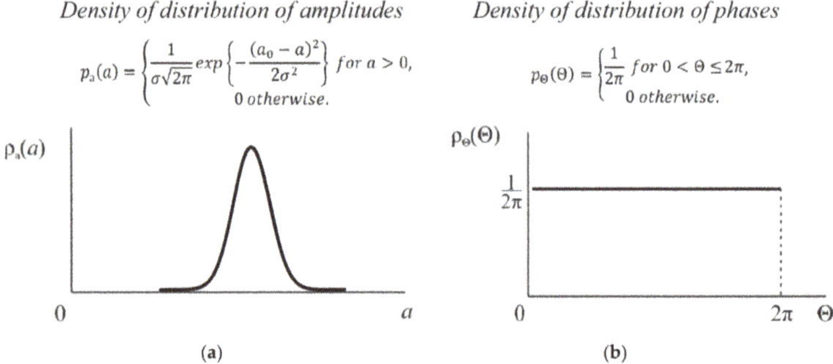

$$p_a(a) = \begin{cases} \dfrac{1}{\sigma\sqrt{2\pi}}\,exp\left\{-\dfrac{(a_0-a)^2}{2\sigma^2}\right\} & for\ a > 0, \\ 0\ otherwise. \end{cases}$$

$$p_\Theta(\theta) = \begin{cases} \dfrac{1}{2\pi} & for\ 0 < \Theta \le 2\pi, \\ 0\ otherwise. \end{cases}$$

(**a**) (**b**)

Figure 2. The density of the amplitude (**a**) and phase (**b**) distributions of backscattered radiation.

The signal level in this device is determined by the parameters of the probing pulse: the peak power and duration. The power from above is limited by the threshold for the appearance of nonlinear effects, primarily modulation instability [17,54], and the duration is limited by the spatial resolution necessary for the device, which is the half-width of the pulse in the fiber.

The reflectogram obtained from the device has a jagged appearance, varying from maximum to minimum, depending on how the interference of backscattered waves occurs in each area of the sensor. The resulting form of the reflectogram remains stable until the scattering centers' positions change. When they are displaced under external influence, the form of the reflectogram changes. On this basis, one can discuss the activity in this area of the sensor.

In the case of a system based on Rayleigh scattering, the signal formation can be represented as a result of the backscattered waves being added from all of the inhomogeneities within the half-width of the pulse duration. Since each wave has a random amplitude with a Gaussian probability density distribution, the change in the phase of each wave due to the scattering center's displacement makes a nonlinear contribution to the change in the final signal recorded by the system. This process is shown in a simplified form in Figure 3, which, for clarity, shows the result of adding 10 waves before (solid line) and during impact Δl on the fiber (dotted line). The colors show the waves from each center of scattering.

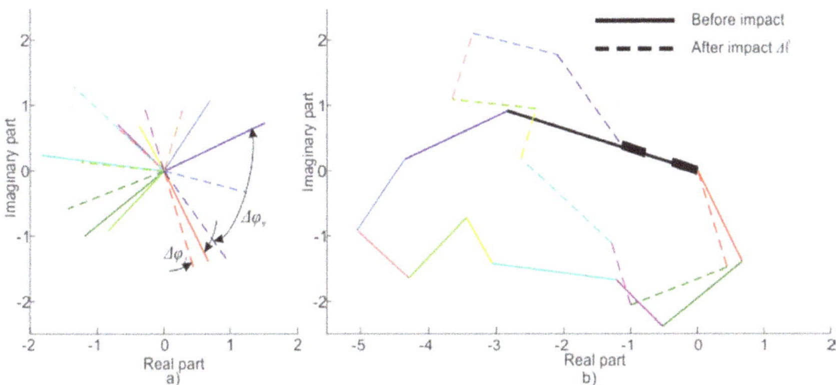

Figure 3. Scheme of signal generation in a phi-OTDR: components (**a**) and summation (**b**). A line of each color represents one scattered wave: the length is equal to the amplitude and the angle is equal to the phase.

Mathematically, this can be expressed as the sum of n backscattered waves from inhomogeneities within the pulse half-width that have a Gaussian distribution of amplitudes

$$p_E(a) = \begin{cases} \dfrac{1}{\sigma\sqrt{2\pi}}\exp\left(-\dfrac{(a-a_0)^2}{2\sigma^2}\right) & for\ a > 0, \\ 0\ otherwise, \end{cases} \tag{1}$$

$$E^2 = \left|\sum_{m=1}^{n} E_m\exp(-kr_m)\right|^2, \tag{2}$$

$$\Delta\varphi = arg\left(\sum_{m=1}^{n} E_m exp\left(-\frac{2\pi l_m}{\lambda}\right)\right) - arg\left(\sum_{m=1}^{n} E_m exp\left(-\frac{2\pi l'_m}{\lambda}\right)\right), \tag{3}$$

where a is the amplitude of the scattering center, a_0 is the expected value of the amplitude, k is the wavenumber of the backscattered wave, p_E is the function describing the probability density of the wave amplitude, and σ is the Gauss distribution parameter. Additionally, E_m is the amplitude of the m-th scattering center out of n included in the pulse half-width, l_m is the distance from the radiation source to the m-th scattering center, λ is the current wavelength of the system's laser source, and $\Delta\varphi$ is the signal phase change caused by fiber deformation.

This shows that the phase changes of each backscattered wave will differ when the fiber section is linearly stretched: all phase changes from the nearer sections will be added to the phase change in the distant sections. As a result, the increase in the phase change will increase for different sections in different ways, and the signal will not change according to a simple sinusoidal law but a complex harmonic law.

This expression shows that the displacement of each center of scattering contributes to the change in the interference result and the signal's magnitude from the sensor area. Yet, it is impossible to predict whether this change will lead to an increase or decrease in the signal or to predict its monotony and periodicity. One of the options for changing the resulting intensity is shown in Figure 4.

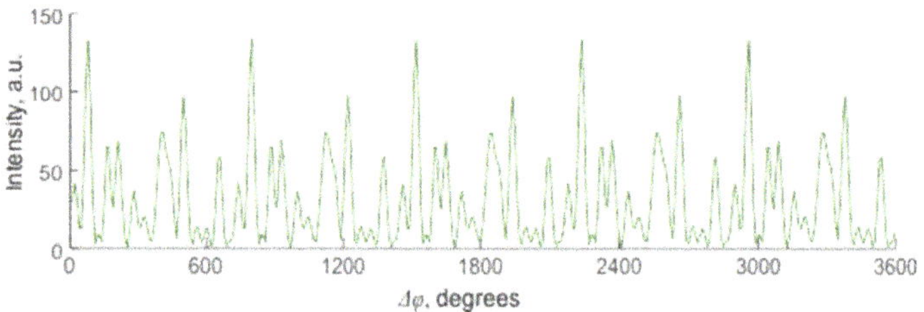

Figure 4. Change in the resulting intensity of the backscattered signal depending on the amount of deformation in a system with many randomly located reflectors.

2.2. Signal Conditioning in a WFBG-Based System

A sensor based on a fiber with periodically applied WFBG can be used to obtain a higher-intensity signal from the sensor. The technology allows creating such structures directly during the extraction of fiber, which makes it convenient and inexpensive [28,30]. The reflection coefficient of such WFBGs can be fractions of a percent or less, while their relative amplitude spread in the array will be no more than 15% [55]. A scheme of the system based on WFBGs is shown in Figure 5. The operating principles are similar to those of a phase-sensitive reflectometer, but the signal back-reflected from the sensor fiber has significantly higher power, and a preamplifier and a filter are not needed to suppress spontaneous emission.

To interpret the received data, it can be assumed that each section between adjacent WFBGs generated a separate signal. Notably, signal interference will be observed only in the case when the pulse length L_{pulse} is twice the distance of that between WFBG1 and WFBG2 L because to observe the interference of reflected signals, the pulse reflected from WFBG2 must have time to return and interact with the signal reflected from WFBG1. Provided that all WFBGs are located at the same distance, then the length of the scanning pulse must meet the following condition to obtain a clear interference pattern from two adjacent WFBGs:

$$2L < L_{pulse} < 4L, \tag{4}$$

In this case, the pulse duration should be less than $4L$, so that reflection from more distant gratings will not affect the interference pattern. The interference of reflected signals from two neighboring WFBGs is demonstrated in Figure 6.

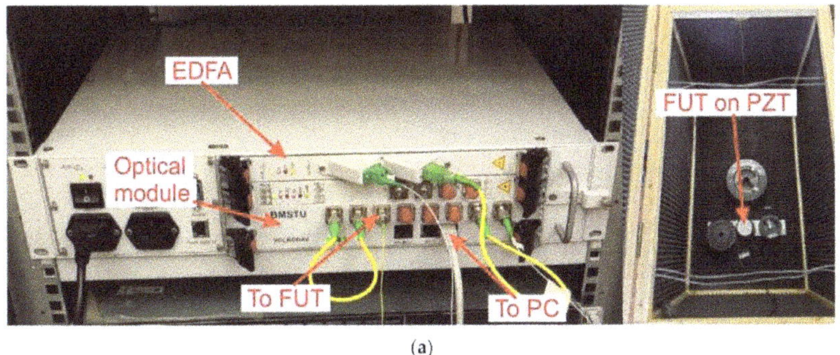

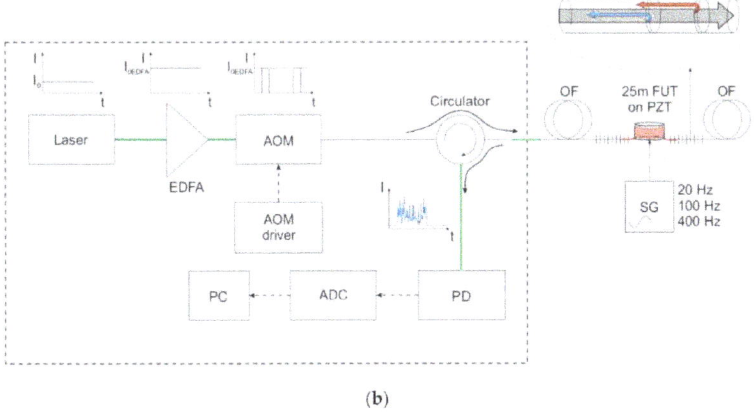

Figure 5. Photo and scheme of a WFBG-based system for experimenting to define sensitivity. Laser: radiation source, EDFA: optical erbium amplifier, OM: optical modulator, OF: optical fiber, FUT: fiber section for testing, PD: photodiode, ADC: analog-to-digital converter, PC: personal computer. Thin solid lines—fiber optic connections inside the optical module. Bold solid lines—fiber connections outside the case. Dashed lines—electrical connections. (**a**): photo of setup; (**b**): scheme of setup.

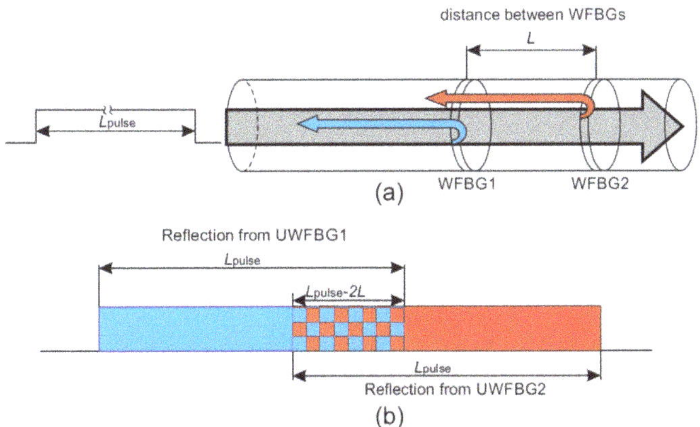

Figure 6. Backscattered signal formation from the WFBG pair. (**a**): Pulse reflection from neighboring WFBGs; (**b**): Scheme of reflected light interference.

Thus, a sensor with periodically applied WFBGs can be compared to a sensor in a conventional reflectometer with scattering centers located with a large and strict periodicity and the same reflection coefficients. In comparing the back-reflected power, a conventional reflectometer, as mentioned earlier, will have a value of $10^{-4}\%$, and a modified one will have a value of R—that is the WFBG reflection coefficient. In other words, the value of the useful signal can be several orders of magnitude higher, which allows it to be registered without an erbium preamplifier. The only limitation on the WFBG reflection level is the number of sections on the scanning line [35,36]. Importantly, when scanning a sensor line, it must be possible to detect signals from distant sections.

Thus, the use of sensors with WFBGs should theoretically reduce the noise level and allow impacts on a sensor with a much lower intensity to be recorded than is possible with conventional fiber. The experimental determination of this quantity using laboratory facilities is described in the next section.

It is also necessary to note the peculiarity of the signal formation from lines with WFBGs in the case of classical phi-OTDR. The difference in deformations is shown in Figure 7. With a uniform stretching of the section for a phi-OTDR, the distance between each pair of adjacent scattering centers increases by concrete value for this pair. For a sensor with WFBGs, the equivalent stretch of the fiber section will affect the phase difference only between the backscattered waves from WFBG_i and WFBG_{i+1}, which makes the dynamics of the fiber-deformation process easy to understand.

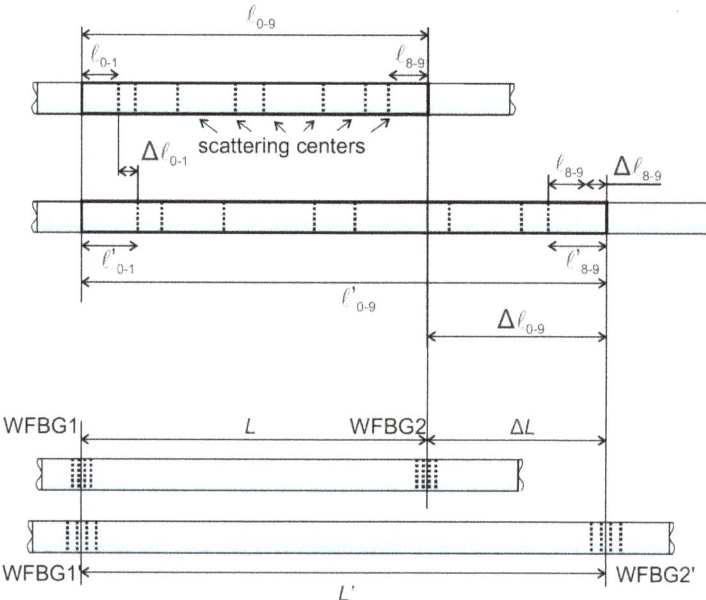

Figure 7. Scheme for changing the distance to the reflector in a phase-sensitive reflectometer based on backward Rayleigh scattering (top) and using WFBGs (bottom).

For WFBGs, interference occurs between two waves reflected from adjacent structures. In this case, any periodic or monotonic fiber deformations between them will change the signal from its minimum to its maximum value. Figure 8 shows reflected waves from the first WFBG1 ($\varphi_1 = \text{const}$) and the second WFBG2 ($\varphi_2 = \text{var}$) as well as the resulting wave. These plots are presented on a complex plane, with the real and imaginary components of the light wave plotted along the abscissa and ordinate, respectively, and the arrows denoting the added backscattered components. Their modulus is equal to the wave amplitude, and the angle with the abscissa axis is the current phase. In this case, several backscattered waves can be added to the resulting vector according to the rules of vector addition. A linear change in length between two adjacent WFBGs results in a harmonic change in the sum signal, as shown in Figure 9, which represents the change over time in the resulting vector's amplitude.

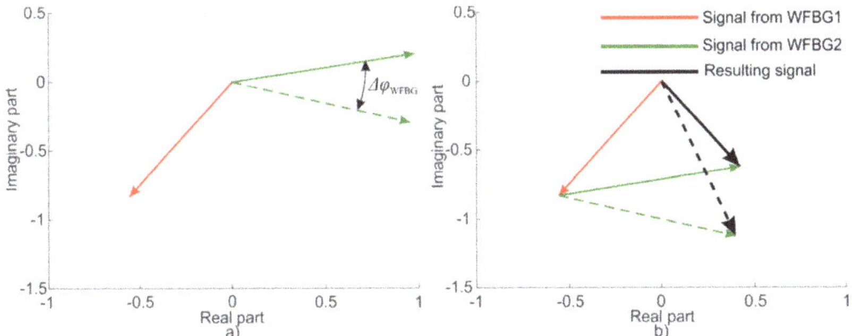

Figure 8. Signal conditioning scheme from two WFBGs: components (**a**) and summation (**b**).

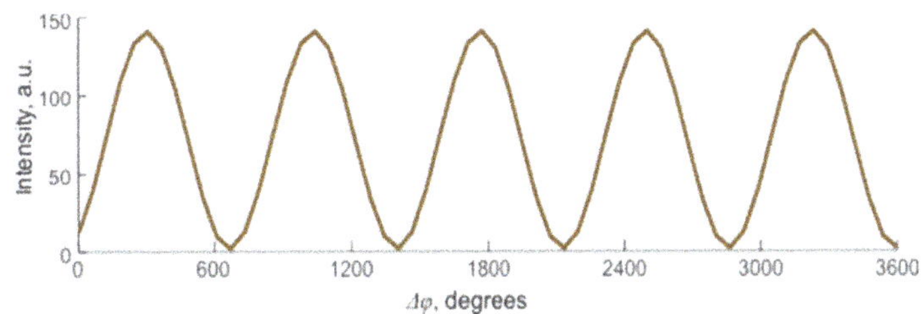

Figure 9. Change in the resulting intensity of the back-reflected signal depending on the distance between adjacent WFBGs.

Mathematically, this can be expressed as the sum of two waves. The amplitude difference can be neglected because it will be no more than 1% between neighboring gratings.

$$E^2 = E_1^2 + E_2^2 + 2E_1 E_2 cos(\varphi_2 - \varphi_1) \tag{5}$$

To describe the formation of a back-reflected signal from neighboring WFBGs, we can take φ_1 as a constant because the deformations in the measured section's fiber will not affect the phase difference of the signal in the section itself. In this case, the phase of the second grating due to the effect changes by the value $\Delta\varphi_{WFBG}$:

$$E^2 = E_1^2 + E_2^2 + 2E_1 E_2 cos((\varphi_2 + \Delta\varphi_{WFBG}) - \varphi_1). \tag{6}$$

This expression shows that with a linear phase change, the system signal will change as a function of the cosine. In this case, the magnitude of the phase change itself will be determined by the overall change in the length of the fiber between the gratings:

$$\Delta\varphi_{WFBG} = \frac{4\pi\Delta L}{\lambda}. \tag{7}$$

Finally, we can formulate common requirements for WFBGs for phi-OTDR interrogation in the Table 1.

Table 1. Common parameter of WFBG for phi-OTDR.

Parameter	WFBG-OTDR
Reflectivity, %	From 0.001 to 1, depending on needed sensor length [35]
Reflectivity instability (relative), %	<15
Central wavelength	Similar to the laser source of the system
Central wavelength instability	Less than 10% of WFBG spectral width
Spectral width, nm	Better, not less than 1 to avoid temperature and strain influence on the reflectivity spectrum. Fiber optic cable which is usually used for sensor installation provides some protection from external damaging influences, but some wavelength shifts due to seasons of the year or slow ground movements are still possible. The wide spectrum of WFBG reduces their effect on reflectivity changes and makes easier the process of laser wavelength and reflectivity spectrum matching.

3. Calculation

3.1. Calculation of the Signal and Noise Level in a Phase-Sensitive Reflectometer

As mentioned above, the main components generating noise in this system are laser wavelength instability, optical preamplifier noise, and photodetector module noise. A detailed analysis of the influences of instability in the laser source's radiation is given in [18,56]. They can only be eliminated when using an ideal laser source or a source with negligible frequency fluctuations. The existing sources are not ideal, but their noise level is low enough not to consider this type of noise as the main one.

The noise of a photodetector module is determined by the quality and technology of its manufacturing. It is proportional to the specific equivalent noise power A and the root of the frequency band Δf recorded by the radiation receiver, which, in turn, is determined by the spatial resolution of the system Δz. Using the average value $\Delta z = 20$ m, attained at a pulse duration $\tau = 200$ ns, the minimum required frequency band recorded and orbited by the photodetector module will be $\Delta f = 1/\tau = 5$ MHz. Higher frequency ADCs are often used when increasing the number of points for processing and detection, from 50 to 100 MHz. This makes it possible to increase the sampling resolution but keeps the spatial resolution the same.

The scheme of phi-OTDR uses a preamplifier generating the third type of noise in question to boost the extremely low backscattered signal level. With a pulse duration $\tau = 200$ ns, the backscattered fraction of the pulse is approximately 10^{-6}, following the expression

$$r = r_1 + 10\log(\tau) = -80 + 10\log(200) = -57 \text{ dB},\tag{8}$$

where $r_1 = -80$ dB—is the power fraction scattered by a 1 ns pulse.

This value is the most frequently encountered when creating monitoring systems for extended objects. Increasing it will give a higher level of received power but will worsen the spatial resolution. The power of the signal coming from the L_S km sensor can be obtained from the expression

$$P_{sig} = P_{in} + \eta + 2\alpha L_S + r,\tag{9}$$

where $P_{in} = 23$ dBm (the upper value is limited by nonlinear effects [17]) is the peak pulse power, $\eta = -4$ dB is the loss at connectors and elements, $\alpha = -0.18$ dB/km is the fiber-attenuation coefficient, and L_S is usually no more than 50 km. That is, P_{sig} ranges from 146 nW (nearby sites–less than 1 km) to 5.8 nW (distant sites–about 40 km).

The magnitude of the generated signal-spontaneous noise can be determined by the expression [53]

$$\sigma_{s-sp}{}^2 = 2G^2 P_{sig} NFh\nu\Delta f,\tag{10}$$

where $G = 1000$ is the preamplifier gain, P_{sig}, nW is the amplified signal power, $NF = 4$ is the preamplifier noise factor, $h = 6.626 \cdot 10^{-34}$ J·s is the Planck constant, $v = 193.4$ THz is the radiation frequency, and $\Delta f = 5$ MHz is the receiver frequency bandwidth. The value of σ_{s-sp} ranges from 0.9 (nearby sections) to 0.2 μW (distant sections). In this case, the signal power will range (after amplification by a factor of G) from 150 (nearby section) to 6 μW (distant section). Note that the preamplifier also generates spontaneous-spontaneous noise σ_{sp-sp}

$$\sigma_{sp-sp}{}^2 = (NFhvG)^2 2\Delta f(\Delta v - \Delta f/2) \approx (36\text{nW})^2, \tag{11}$$

where Δv is the optical filter's bandwidth. When using better and more expensive samples with $\Delta v = 1$ GHz, this value is two orders of magnitude less than the beat noise of signal radiation with spontaneous radiation. This value will be higher when using the more common and less expensive filters with $\Delta v = 12.5$ GHz, but still will have an order of magnitude less signal-spontaneous noise. It is impossible to not use a preamplifier in the standard scheme. Without one, only signals from the first several kilometers of the sensor could be measured, with the help of existing radiation receivers. The photodetector's noise level is determined by the frequency band and the equivalent noise power (a receiver characteristic), and its value is similar to that of the level of spontaneous-spontaneous noise $\sigma_{pd} = 10$ nW.

3.2. Calculation of the Signal and Noise Levels in the WFBG System

The WFRB system does not use an optical preamplifier, and the main parameter for calculating the signal power is the gratings' reflection coefficient. The expression can be written as follows

$$P_{WFBGsig} = P_{in}(1 - R)^{2(N-1)}exp(-2\alpha_{WFBG}L_S)R, \tag{12}$$

where P_{in}, W is the peak power, α_{WFBG}, km^{-1} is the fiber attenuation with WFBGs, N is the number of gratings per 1 km, L, km is the sensor length, and R is the reflectance of one WFBG.

In this case, the peak pulse power is not limited by nonlinear effects—since the gratings' reflection coefficient is orders of magnitude higher than the Rayleigh scattering value—but by the dynamic range of the receiving part's intensity, in particular the ADC. Studies were carried out on the same components for both the phi-OTDR and the WFBG system. Therefore, the parameters of the peak pulse power were selected to fall into the same power range. The maximum (from the nearby site) was $P_{WFBGsig} = 150$ μW. For this, given the assembled experimental scheme based on a sensor fiber with parameters $R = 0.003$, $N = 50$, it was necessary to inject 70 mW pulses into the sensor following Equation (12). In this case, the main source of the noise was the photodetector or the laser source's instability, as in the experiment with a phase-sensitive reflectometer, $\sigma_{pd} = 10$ nW. The effect of shifting the radiation wavelength from λ to $\lambda + \Delta\lambda$ can be estimated from Expression (6)

$$\Delta\varphi_{WFBG} = \frac{4\pi L}{\lambda} - \frac{4\pi L}{\lambda + \Delta\lambda} = 4\pi L \frac{\Delta\lambda}{\lambda(\lambda + \Delta\lambda)} \approx 4\pi L \frac{\Delta\lambda}{\lambda^2} = 4\pi L \frac{\Delta v_{las}}{c}, \tag{13}$$

where L is the distance between the gratings, c is the speed of light, and Δv_{las} is the width of the laser line. We used a BasikX15 laser with a linewidth of less than 300 Hz and a grating spacing of 20 m, which provided an estimated phase shift of less than 0.085 mrad—that is, less than 0.003% of the maximum intensity fluctuation value of π rad. The noise level created by the laser source's frequency instability will be less than $\sigma_{las} = 0.003\%(P_{WFBGsig}) = 4.5$ nW, which, in this case, is comparable to the receiver's noise level. However, note that many laser sources have a wider bandwidth, and this type of noise will become the main one in the system.

In this way, the increase in signal-to-noise ratio can be estimated based on the decreasing noise level while the emission carrier's power is maintained. This value will be

$$SNR_{enh} = \frac{\sigma_{s-sp}}{\sqrt{\sigma_{pd}{}^2 + \sigma_{las}{}^2}} = \frac{900\,\text{nW}}{11\,\text{nW}} = 82 \ times \approx 19 \ \text{dB}. \tag{14}$$

However, the useful signal from sensors of this type is not the general power level but its deviation when the effect appears. In addition, because the signal itself and the response to it are non-stationary in time, it is easier to obtain information based on a statistical characteristic, specifically the RMS of the signal section's intensity in the time interval corresponding to the effect on the fiber. The ratio of the RMS of the section with the signal to that of the section without the signal is taken to determine the signal-to-noise ratio.

4. Results

For the experiments, two setups were assembled: a phase-sensitive reflectometer utilizing backward Rayleigh scattering and a schematic of a WFBG reflectometer with a WFBG reflection coefficient $R = 3 \times 10^{-3}$ and $N = 50$ gratings per km (Figures 1 and 5, respectively). We were interested in SNR enhancement from the nearby site (<1 km). The values for other distances and periodicities of the gratings' arrangement can be determined using the calculations in [35,36].

The amplitude of the recorded signal was tested using cylindrical piezoelectrics, on which a sensor fiber was wound in one layer without overlaps. For a conventional phi-OTDR, the length of the wound cable was 22 m. This distance does not violate the equivalence to the sensitive area of the WFBG-based system, which is equal to the distance between the gratings (20 m) since the resolution of the reflectometric system is determined only by the pulse duration $\Delta z = c\tau/(2n)$, and at $\tau = 200$ ns, it will also be 20 m. This value makes it possible to guarantee that at least one segment, which is entirely interrogated by the probe pulse, will fall into the phi-OTDR trace, almost without increasing the area of influence on the fiber. The pulse parameters are given in Table 2. For a WFBG-based OTDR, a section between adjacent WFBGs was wound on a piezoelectric cylinder. The pulse length was chosen to guarantee the interference of this pair of gratings without the neighboring ones being captured.

Table 2. Parameters of experimental setups.

Parameter	φ-OTDR	wFBG-OTDR
Resolution, m	20 (determined by the duration of the probe pulse)	20 (determined by the distance between the gratings)
Fiber length at PZT, m	22	20
Fiber strain range at PZT, nm	From 10 to 5000	From 5 to 100
Range of supplied frequencies, Hz	20, 100, 400	20, 100, 400

The nearby section of the fiber (within the first kilometer) was tested. The power of the optical amplifiers was adjusted to obtain a signal in the tested section at a peak level of 150 µW, as used in the previous calculations. The main goal was to analyze the magnitude of the improvement in the signal-to-noise ratio in this section when switching from a phase-sensitive reflectometer to a system based on WFBGs. The transition to the calculated sensitivities in the distant sections of the sensor can be found based on the expressions given in the calculation part of this work.

The signals received from the installations are plotted in Figure 10 for φ-OTDR and in Figure 11 for WFBG-OTDR. The impact was carried out with different amplitudes at a 20 Hz frequency. The maximum strain values are shown on the graph opposite the corresponding system response.

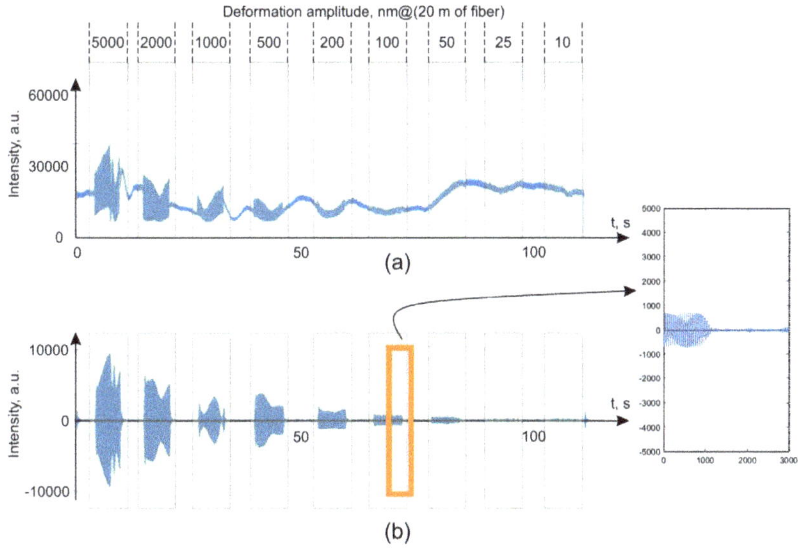

Figure 10. Initial data from the signal of a phase-sensitive reflectometer utilizing backward Rayleigh scattering when exposed to 20 Hz (**a**) and filtered for frequencies of 18–22 Hz (**b**) with zoomed data for 100 nm amplitude impact.

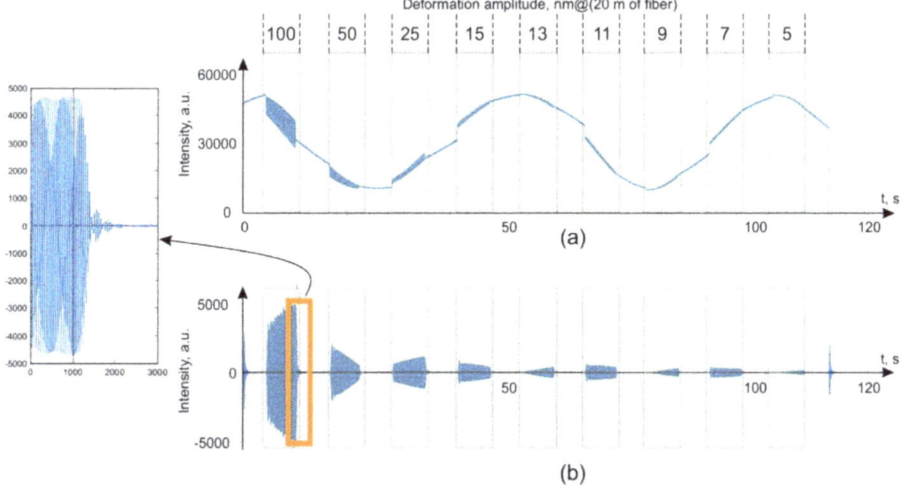

Figure 11. Initial data from the signal of a reflectometer based on WFBG with a pulse duration of 300 ns (provides spatial resolution 20 m when the distance between WFBGs is equal 20 m) when exposed to 20 Hz (**a**) and filtered for frequencies of 18–22 Hz (**b**) with zoomed data for 100 nm amplitude impact.

The presented graphs show the main differences in the general forms of the signals from the backward Rayleigh scattering reflectometers versus those based on WFBGs. The signal from the WFBGs changes smoothly, in the form of a sinusoidal curve, due to the monotonous temperature deviation of the sensor, as shown in the theory section. The signal from a backward Rayleigh scattering reflectometer has a complicated envelope due to the interference of backscattered waves from a huge number of scattering centers randomly located in the fiber. For the filtered signals, the signal-to-noise ratio was determined as the ratio of the root–mean–square deviation (RMS) of the area's signal to the impact of the RMSD for the signal in the unimpacted area. A comparison of RMSD values is more useful for detection algorithms applications [57–59] than a comparison of spectral peak's level [46]. The data were filtered with a second-order Butterworth bandpass filter in the range from 18 to 22 Hz. In this case, all of the sensor readings were passed through the filter in the range of ±50 m from the coordinates at which the action was generated, after which the sample was selected at which the deflection caused by fiber deformation had the largest maximum amplitude. The readout from the impact coordinates deviated by no more than ±5 m.

The frequency of 20 Hz was chosen because it is one of the most important frequencies for recording vibration signals in the ground [60,61]. The following values of the signal-to-noise ratios were obtained for this frequency, as presented in the graph in Figure 12a.

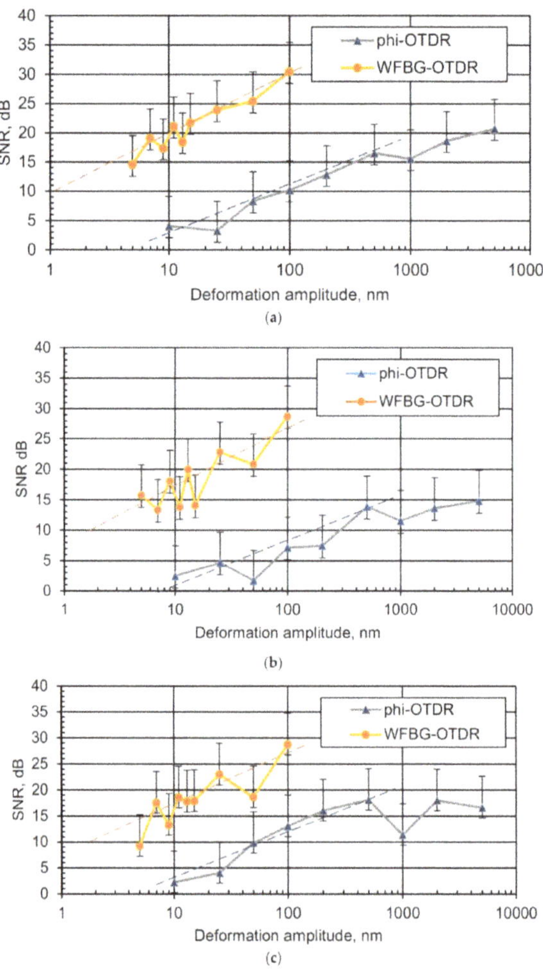

Figure 12. Graphs of the SNR for OTDRs using backscattering and based on WFBGs when recording the impacts on the fiber with different deformation amplitudes at (**a**) 20 Hz, (**b**) 100 Hz, and (**c**) 400 Hz frequencies. Dashed lines—trends of SNR dependence from amplitude.

The difference between the graphs is slightly less than two orders of magnitude, which is consistent with the previously calculated value SNR_{enh}. The sensitivity study for two other frequencies—100 Hz (the upper limit of the propagated waves in ordinary soil media) and 400 Hz (vibrational waves propagating in ice and rocks)—are shown in Figure 12b,c, respectively.

As in the case of exposure at a frequency of 20 Hz, the WFBG-based system also shows a higher signal-to-noise ratio at frequencies of 100 and 400 Hz, and therefore greater sensitivity, by about 1.5–2 orders of magnitude, as compared to the classical phi-OTDR scheme. Trend lines show that the 10 dB SNR threshold for the WFBG-OTDR system is located in the impact range from 1 to 2 nm, which is also 1.5–2 orders better than the phi-OTDR scheme. For phi-OTDR, without phase reconstruction, SNR growth stops near 1000 nm impact amplitude due to signal wrapping.

5. Discussion

In this work, the signal formation was analyzed in a phase-sensitive reflectometer and a system based on WFBGs. On this basis, the main noise components in each scheme were determined, and the theoretical value of the sensitivity increase in the WFBG system was calculated. In our calculations, we used the assumption that the recorded return signal should be into the same range in intensity for phi-OTDR and WFBG-OTDR systems, and only the system's noise is present without background external influences on fiber. This was proposed to calculate the near zone of the cable, where the attenuation in the sensor fiber can be neglected. These assumptions allow an estimate of the signal-to-noise improvement for fiber with WFBG. The reduction in the desired signal from attenuation can be performed for each specific fiber based on its parameters. The experimental results show that the calculated values agree within the error limits, mainly due to the random nature of scattering in the sensor of a phase-sensitive reflectometer. For a test fiber sample with basic parameters R = 0.003, N = 50 gratings per km, the total noise decreased by about 19 dB, which allows a WFBG-OTDR to register weaker or distant impacts on the sensor.

Author Contributions: Conceptualization, A.B.P. and O.V.B.; methodology, K.V.S.; software, K.V.S.; validation, A.A.Z., A.O.C., and K.I.K.; formal analysis, A.I.L., and A.A.Z.; investigation, K.V.S., and K.I.K.; resources, A.I.L.; data curation, A.A.Z.; writing—original draft preparation, K.V.S.; writing—review and editing, A.A.Z. and O.V.B.; visualization, A.O.C. and K.I.K.; supervision, O.V.B.; project administration, A.B.P. All authors have read and agreed to the published version of the manuscript.

Funding: This research was carried out within the framework of the state task. Anton Chernutsky was funded by the Russian Foundation of Basic Research, project number 19-32-90185.

Conflicts of Interest: The authors declare no conflict of interest. The funders had no role in the design of the study; in the collection, analyses, or interpretation of data; in the writing of the manuscript, or in the decision to publish the results.

References

1. Taylor, H.F.; Lee, C.E. Apparatus and Method for Fiber Optic Intrusion Sensing. U.S. Patent 5194847, 16 March 1993.
2. Juarez, J.C.; Maier, E.W.; Choi, K.N.; Taylor, H.F. Distributed fiber-optic intrusion sensor system. *J. Lightwave Technol.* **2005**, *23*, 2081–2087. [CrossRef]
3. Juarez, J.C.; Taylor, H.F. Field test of a distributed fiber-optic intrusion sensor system for long perimeters. *Appl. Opt.* **2007**, *46*, 1968–1971. [CrossRef] [PubMed]
4. Choi, K.N.; Juarez, J.C.; Taylor, H.F. Distributed fiberoptic pressureseismic sensor for low-cost monitoring of long perimeters. *Proc. SPIE* **2003**, *5090*, 134–141. [CrossRef]
5. Park, J.; Lee, W.; Taylor, H.F. A fiber optic intrusion sensor with the configuration of an optical time domain reflectometer using coherent interference of Rayleigh backscattering. *Proc. SPIE* **1998**, *3555*, 49–56. [CrossRef]
6. Lu, Y.; Zhu, T.; Chen, L.; Bao, X. Distributed vibration sensor based on coherent detection of phase-OTDR. *J. Lightwave Technol.* **2010**, *28*, 3243–3249. [CrossRef]
7. Qin, Z.; Chen, L.; Bao, X. Continuous wavelet transform for non-stationary vibration detection with phase-OTDR. *Opt. Express* **2017**, *20*, 20459–20465. [CrossRef] [PubMed]

8.   Zhang, Z.; Bao, X. Distributed optical fiber vibration sensor based on spectrum analysis of Polarization-OTDR system. *Opt. Express* **2008**, *16*, 10240–10247. [CrossRef] [PubMed]

9.   Tejedor, J.; Macias-Guarasa, J.; Martins, H.F.; Piote, D.; Pastor-Graells, J.; Martin-Lopez, S.; Corredera, P.; Gonzalez-Herraez, M. A novel fiber optic based surveillance system for prevention of pipeline integrity threats. *Sensors* **2017**, *17*, 355. [CrossRef]

10. Tejedor, J.; Macias-Guarasa, J.; Martins, H.F.; Pastor-Graells, J.; Martín-López, S.; Guillén, P.C.; Pauw, G.D.; Smet, F.D.; Postvoll, W.; Ahlen, C.H.; et al. Real field deployment of a smart fiber-optic surveillance system for pipeline integrity threat detection: Architectural issues and blind field test results. *J. Lightwave Technol.* **2018**, *36*, 1052–1062. [CrossRef]

11. Pastor-Graells, J.; Martins, H.F.; Garcia-Ruiz, A.; Martin-Lopez, S.; Gonzalez-Herraez, M. Single-shot distributed temperature and strain tracking using direct detection phase-sensitive OTDR with chirped pulses. *Opt. Express* **2016**, *24*, 13121–13133. [CrossRef]

12. Pastor-Graells, J.; Nuno, J.; Fernandez-Ruiz, M.R.; Garcia-Ruiz, A.; Martins, H.F.; Martin-Lopez, S.; Gonzalez-Herraez, M. Chirped-Pulse Phase-Sensitive Reflectometer Assisted by First -Order Raman Amplification. *J. Lightwave Technol.* **2017**, *35*, 4677–4683. [CrossRef]

13. Shatalin, S.V.; Treschikov, V.N.; Rogers, A.J. Interferometric optical time-domain reflectometry for distributed optical-fiber sensing. *Appl. Opt.* **1998**, *37*, 5600–5604. [CrossRef]

14. Alekseev, A.E.; Vdovenko, V.S.; Gorshkov, B.G.; Potapov, V.T.; Simikin, D.E. A phase-sensitive optical time-domain reflectometer with dual-pulse phase modulated probe signal. *Laser Phys.* **2014**, *24*, 115106. [CrossRef]

15. Alekseev, A.E.; Vdovenko, V.S.; Gorshkov, B.G.; Potapov, V.T.; Simikin, D.E. A phase-sensitive optical time-domain reflectometer with dual-pulse diverse frequency probe signal. *Laser Phys.* **2015**, *25*, 065101. [CrossRef]

16. Tosoni, O.; Aksenov, S.B.; Podivilov, E.V.; Babin, S.A. Model of a fibreoptic phase-sensitive reflectometer and its comparison with the experiment. *Quantum Electron.* **2010**, *40*, 887–892. [CrossRef]

17. Nesterov, E.T.; Zhirnov, A.A.; Stepanov, K.V.; Pnev, A.B.; Karasik, V.E.; Tezadov, Y.A.; Kondrashin, E.V.; Ushakov, A.B. Experimental study of influence of nonlinear effects on phase- sensitive optical time-domain reflectometer operating range. *J. Phys. Conf. Ser.* **2015**, *584*, 012028. [CrossRef]

18. Zhirnov, A.A.; Stepanov, K.V.; Chernutsky, A.O.; Fedorov, A.K.; Nesterov, E.T.; Svelto, C.; Pnev, A.B.; Karasik, V.E. Influence of the Laser Frequency Drift in Phase-Sensitive Optical Time Domain Reflectometry. *Opt. Spectrosc.* **2019**, *127*, 656–663. [CrossRef]

19. Nikitin, S.P.; Kuzmenkov, A.I.; Gorbulenko, V.V.; Nanii, O.E.; Treshchikov, V.N. Distributed temperature sensor based on a phase-sensitive optical time-domain Rayleigh reflectometer. *Laser Phys.* **2018**, *28*, 085107. [CrossRef]

20. Yatseev, V.A.; Zotov, A.M.; Butov, O.V. Use of a chirped pulse for restoring the phase in a coherent reflectometer. *Foton Express* **2019**, *6*, 46–47.

21. Hartog, A.H.; Liokumovich, L.B.; Ushakov, N.A.; Kotov, O.I.; Dean, T.; Cuny, T.; Constantinou, A.; Englich, F.V. The use of multi-frequency acquisition to significantly improve the quality of fibre-optic-distributed vibration sensing. *Geophys. Prospect.* **2018**, *66*, 192–202. [CrossRef]

22. Hartog, A.H.; Liokumovich, L.B. Phase Sensitive Coherent otdr with Multi-Frequency Interrogation. U.S. Patent WO2013066654 A1, 22 October 2012.

23. Zhu, F.; Zhang, X.; Xia, L.; Guo, Z.; Zhang, Y. Active compensation method for light source frequency drifting in Phi-OTDR sensing system. *IEEE Photon. Technol. Lett.* **2015**, *27*, 2523–2526. [CrossRef]

24. Wu, H.; Yang, M.; Yang, S.; Lu, H.; Wang, C.; Rao, Y. A novel DAS signal recognition method based on spatiotemporal information extraction with 1DCNNs-BiLSTM network. *IEEE Access* **2020**, *8*, 119448–119457. [CrossRef]

25. Tu, G.; Zhang, X.; Zhang, Y.; Zhu, F.; Xia, L.; Nakarmi, B. The development of an phi-OTDR system for quantitative vibration measurement. *IEEE Photon. Technol. Lett.* **2015**, *27*, 1349–1352. [CrossRef]

26. Zhou, L.; Wang, F.; Wang, X.; Pan, Y.; Sun, Z.; Hua, J.; Zhang, X. Distributed Strain and Vibration Sensing System Based on Phase-Sensitive OTDR. *IEEE Photon. Technol. Lett.* **2015**, *27*, 1884–1887. [CrossRef]

27. Kersey, A.D.; Dandridge, A.; Vohra, S.T. Coherent Reflectometric Fiber Bragg Grating Sensor Array. U.S. Patent US6285806B1, 4 September 2001.

28. Popov, S.M.; Butov, O.V.; Kolosovskiy, A.O.; Voloshin, V.V.; Vorob'ev, I.L.; Vyatkin, M.Y.; Fotiadi, A.A.; Chamorovskiy, Y.K. Optical fibres with arrays of FBG: Properties and application. In Proceedings of the 2017 Progress in Electromagnetics Research Symposium–Spring (PIERS), St. Petersburg, Russia, 22–25 May 2017; pp. 1568–1573. [CrossRef]

29. Guo, H.; Liu, F.; Yuan, Y.; Yu, H.; Yang, M. Ultra-weak FBG and its refractive index distribution in the drawing optical fiber. *Opt. Express* **2015**, *23*, 4829–4838. [CrossRef] [PubMed]

30. Zaitsev, I.A.; Butov, O.V.; Voloshin, V.V.; Vorob'ev, I.L.; Vyatkin, M.Y.; Kolosovskii, A.O.; Popov, S.M.; Chamorovskii, Y.K. Optical Fiber with Distributed Bragg-Type Reflector. *J. Commun. Technol. Electron.* **2016**, *61*, 639–645. [CrossRef]

31. Wang, Y.M.; Gong, J.M.; Wang, D.Y.; Dong, B.; Bi, W.; Wang, A. A quasi-distributed sensing network with time-division-multiplexed fiber Bragg gratings. *IEEE Photon. Technol. Lett.* **2011**, *23*, 70–72. [CrossRef]

32. Hu, C.Y.; Wen, H.Q.; Bai, W. A Novel Interrogation System for Large Scale Sensing Network with Identical Ultra-Weak Fiber Bragg Gratings. *J. Lightwave Technol.* **2014**, *32*, 1406–1411. [CrossRef]

33. Wang, C.; Shang, Y.; Liu, X.H.; Wang, C.; Yu, H.H.; Jiang, D.S.; Peng, G.D. Distributed OTDR-interferometric sensing network with identical ultra-weak fiber bragg gratings. *Opt. Express* **2015**, *23*, 29038–29046. [CrossRef]

34. Wang, X.C.; Yan, Z.J.; Wang, F.; Sun, Z.Y.; Mou, C.B.; Zhang, X.P.; Zhang, L. SNR enhanced distributed vibration fiber sensing system employing polarization OTDR and ultraweak FBGs. *IEEE Photon. J.* **2015**, *7*, 6800511. [CrossRef]

35. Xia, L.; Zhang, Y.; Zhu, F.; Cao, C.; Zhang, X. The performance limit of Φ-OTDR sensing system enhanced with ultra-weak fiber Bragg grating array. *Proc. SPIE* **2015**, *9620*, 962003. [CrossRef]

36. Zhu, F.; Zhang, Y.; Xia, L.; Wu, X.; Zhang, X. Improved Φ-OTDR sensing system for high-precision dynamic strain measurement based on ultra-weak fiber bragg grating array. *J. Lightwave Technol.* **2015**, *33*, 4775–4780. [CrossRef]

37. Zhang, X.; Guo, Z.; Shan, Y.; Sun, Z.; Fu, S.; Zhang, Y. Enhanced Φ-OTDR system for quantitative strain measurement based on ultra-weak fiber Bragg grating array. *Opt. Eng.* **2016**, *55*, 054103. [CrossRef]

38. Zhang, X.; Sun, Z.; Shan, Y.; Li, Y.; Wang, F.; Zeng, J.; Zhang, Y. A high performance distributed optical fiber sensor based on Φ-OTDR for dynamic strain measurement. *IEEE Photon. J.* **2017**, *9*, 1–12. [CrossRef]

39. Li, W.; Zhang, J. Distributed weak fiber Bragg grating vibration sensing system based on 3 × 3 fiber coupler. *Photon. Sens.* **2018**, *8*, 146–156. [CrossRef]

40. Li, Z.; Tong, Y.; Fu, X.; Wang, J.; Guo, Q.; Yu, H.; Bao, X. Simultaneous distributed static and dynamic sensing based on ultra-short fiber Bragg gratings. *Opt. Express* **2018**, *26*, 17437–17446. [CrossRef]

41. Liu, T.; Wang, F.; Zhang, X.; Yuan, Q.; Niu, J.; Zhang, L.; Wei, T. Interrogation of ultra-weak FBG array using double-pulse and heterodyne detection. *IEEE Photon. Technol. Lett.* **2018**, *30*, 677–680. [CrossRef]

42. Tang, J.; Li, L.; Guo, H.; Yu, H.; Wen, H.; Yang, M. Distributed acoustic sensing system based on continuous wide-band ultra-weak fiber Bragg grating array. In Proceedings of the 2017 25th Optical Fiber Sensors Conference (OFS), Jeju, Korea, 24–28 April 2017; pp. 1–4. [CrossRef]

43. Shan, Y.; Ji, W.; Dong, X.; Cao, L.; Zabihi, M.; Wang, Q.; Zhang, Y.; Zhang, X. An Enhanced Distributed Acoustic Sensor Based on UWFBG and Self-Heterodyne Detection. *J. Lightwave Technol.* **2019**, *37*, 2700–2705. [CrossRef]

44. Liu, T.; Li, H.; Ai, F.; Wang, J.; Fan, C.; Luo, Y.; Yan, Z.; Liu, D.; Sun, Q. Ultra-high Resolution Distributed Strain Sensing based on Phase-OTDR. In Proceedings of the 2019 Optical Fiber Communications Conference and Exhibition (OFC), San Diego, CA, USA, USA, 3–7 March 2019; pp. 1–3.

45. Zhang, Y.X.; Fu, S.Y.; Chen, Y.S.; Ding, Z.W.; Shan, Y.Y.; Wang, F.; Chen, M.M.; Zhang, X.P.; Meng, Z. A visibility enhanced broadband phase-sensitive OTDR based on the UWFBG array and frequency-division-multiplexing. *Opt. Fiber Technol.* **2019**, *53*, 101995. [CrossRef]

46. Lee, X.; Che, Q.; Liu, X.; Zhu, P.; Wen, H. Effects of weak fiber Bragg gratings on a distributed vibration sensing system based on phase-sensitive optical time-domain reflectometry. *Opt. Eng.* **2019**, *58*, 087103. [CrossRef]

47. Gan, W.; Li, S.; Li, Z.; Sun, L. Identification of ground intrusion in underground structures based on distributed structural vibration detected by ultra-weak FBG sensing technology. *Sensors* **2019**, *19*, 2160. [CrossRef] [PubMed]

48. Hicke, K.; Eisermann, R.; Chruscicki, S. Enhanced distributed fiber optic vibration sensing and simultaneous temperature gradient sensing using traditional C-OTDR and structured fiber with scattering dots. *Sensors* **2019**, *19*, 4114. [CrossRef] [PubMed]

49. Yatseev, V.A.; Zotov, A.M.; Butov, O.V. Combined Frequency and Phase domain time-gated reflectometry based on a fiber with reflection points for absolute measurements. *Results Phys.* **2020**, *19*, 103485. [CrossRef]

50. Wang, Z.; Lu, B.; Zheng, H.; Ye, Q.; Pan, Z.; Cai, H.; Qu, R.; Fang, Z.; Zhao, H. Novel railway-subgrade vibration monitoring technology using phase-sensitive OTDR. In Proceedings of the 2017 25th Optical Fiber Sensors Conference (OFS), Jeju, Korea, 24–28 April 2017; pp. 1–4. [CrossRef]

51. FiberPatrol FP1150. Available online: https://senstar.com/products/buried-sensors/fiberpatrol-fp1150-for-pipeline-tpi/ (accessed on 23 October 2020).

52. Perimeter Intrusion Detection and Security. Available online: https://www.optasense.com/wp-content/uploads/2017/02/Perimeter-Intrusion-Detection-and-Security_Brochure_A4_Digital.pdf (accessed on 23 October 2020).

53. Desurvire, E. *Erbium-Doped Fiber Amplifiers: Principle and Applications*; John Wiley and Sons. Inc.: New York, NY, USA, 1994.

54. Martins, H.F.; Martin-Lopez, S.; Corredera, P.; Salgado, P.; Frazão, O.; González-Herráez, M. Modulation instability-induced fading in phase-sensitive optical time-domain reflectometry. *Opt. Lett.* **2013**, *38*, 872–874. [CrossRef]

55. Popov, S.M.; Butov, O.V.; Kolosovskii, A.O.; Voloshin, V.V.; Vorob'ev, I.L.; Isaev, V.A.; Vyatkin, M.Y.; Fotiadi, A.A.; Chamorovsky, Y.K. Optical fibres and fibre tapers with an array of Bragg gratings. *Quantum Electron.* **2019**, *49*, 1127–1131. [CrossRef]

56. Zhong, X.; Zhang, C.; Li, L.; Liang, S.; Li, Q.; Lü, Q.; Ding, X.; Cao, Q. Influences of laser source on phase sensitivity optical time-domain reflectometer-based distributed intrusion sensor. *Appl. Opt.* **2014**, *53*, 4645–4650. [CrossRef]

57. Kowarik, S.; Hussels, M.T.; Chruscicki, S.; Münzenberger, S.; Lämmerhirt, A.; Pohl, P.; Schubert, M. Fiber Optic Train Monitoring with Distributed Acoustic Sensing: Conventional and Neural Network Data Analysis. *Sensors* **2020**, *20*, 450. [CrossRef]

58. Fedorov, A.K.; Anufriev, M.N.; Zhirnov, A.A.; Stepanov, K.V.; Nesterov, E.T.; Namiot, D.E.; Karasik, V.E.; Pnev, A.B. Note: Gaussian mixture model for event recognition in optical time-domain reflectometry based sensing systems. *Rev. Sci. Instrum.* **2016**, *87*, 036107. [CrossRef]

59. Shi, Y.; Wang, Y.; Zhao, L.; Fan, Z. An event recognition method for Φ-otdr sensing system based on deep learning. *Sensors* **2019**, *19*, 3421. [CrossRef]

60. Kibblewhite, A.C. Attenuation of sound in marine sediments: A review with emphasis on new low-frequency data. *J. Acoust. Soc. Am.* **1989**, *86*, 716–738. [CrossRef]

61. Taherzadeh, S.; Attenborough, K. Deduction of ground impedance from measurements of excess attenuation spectra. *J. Acoust. Soc. Am.* **1999**, *105*, 2039–2042. [CrossRef]

Publisher's Note: MDPI stays neutral with regard to jurisdictional claims in published maps and institutional affiliations.

Review

Ultrafast Fiber Lasers with Low-Dimensional Saturable Absorbers: Status and Prospects

Pulak Chandra Debnath [1,2] and **Dong-Il Yeom** [1,2,*]

1   Department of Energy Systems Research, Ajou University, 206 Worldcup-ro, Yeongtong-gu, Suwon 16499, Korea; pcd2k1@ajou.ac.kr
2   Department of Physics, Ajou University, 206 Worldcup-ro, Yeongtong-gu, Suwon 16499, Korea
*   Correspondence: diyeom@ajou.ac.kr; Tel.: +82-31-219-1937

Abstract: Wide-spectral saturable absorption (SA) in low-dimensional (LD) nanomaterials such as zero-, one-, and two-dimensional materials has been proven experimentally with outstanding results, including low saturation intensity, deep modulation depth, and fast carrier recovery time. LD nanomaterials can therefore be used as SAs for mode-locking or Q-switching to generate ultrafast fiber laser pulses with a high repetition rate and short duration in the visible, near-infrared, and mid-infrared wavelength regions. Here, we review the recent development of emerging LD nanomaterials as SAs for ultrafast mode-locked fiber laser applications in different dispersion regimes such as anomalous and normal dispersion regimes of the laser cavity operating in the near-infrared region, especially at ~1550 nm. The preparation methods, nonlinear optical properties of LD SAs, and various integration schemes for incorporating LD SAs into fiber laser systems are introduced. In addition to these, externally (electrically or optically) controlled pulsed fiber laser behavior and other characteristics of various LD SAs are summarized. Finally, the perspectives and challenges facing LD SA-based mode-locked ultrafast fiber lasers are highlighted.

Keywords: ultrafast fiber laser; saturable absorber; low-dimensional materials; optically/electrically controlled fiber lasers

Citation: Debnath, P.C.; Yeom, D.-I. Ultrafast Fiber Lasers with Low-Dimensional Saturable Absorbers: Status and Prospects. *Sensors* **2021**, *21*, 3676. https://doi.org/10.3390/s21113676

Academic Editors: Min Yong Jeon and Krzysztof Abramski

Received: 13 April 2021
Accepted: 16 May 2021
Published: 25 May 2021

1. Introduction

Ultrafast lasers have been proven as one of the most effective tools for a wide variety of applications in ultra-precision manufacturing, strong-field physics, nonlinear optics, medical diagnosis, astronomical detection, precision measurement, and fundamental scientific research because of their extremely narrow (femtosecond scale) pulse duration and large peak-power [1–6]. Among ultrafast lasers, the passively mode-locked ultrafast fiber laser (MLFL) based on a saturable absorber (SA) has also emerged as one of the most powerful strategies to develop ultrashort pulses (<100 fs) because of their benefits of high beam quality, low cost, efficient structure, alignment-free compact design, and excellent compatibility [6–9]. Passive mode-locking is a technique that creates a preferred environment for pulsed operation of a laser, effectively employing nonlinear polarization rotation (NPR), nonlinear amplifying loop mirror (NALM), and SA techniques [10–13]. In particular, the advancement of SA design is principally based on the evolution of materials having SA behaviors. In the recent past, the most extensively used form of SAs includes the semiconducting saturable absorber mirror (SESAM), a semiconducting quantum well structure prepared by a deposition method named molecular beam epitaxy (MBE) [14–19]. Modulation depth, SA coefficient, saturation fluence, and recovery time are some of the distinguishing characteristics of a SESAM, determined by adjusting the SESAM's structure. The high stability of the SESAM turns it into one of the significant choices as SAs. However, several detrimental features, including prolonged recovery time (~pico-second level), ultra-narrow working wavelength range, low damage threshold, complicated fabrication process, and high cost of SESAM, have guided the scientific community to find new SA materials which can replace SESAM quantum well. The first condition as alternative SA material is to exhibit a nonlinear absorption behavior such that the optical transmittance efficiently increases as the input laser power increases. Other requirements include a high damage threshold, wide operating range, fast recovery time, low cost, and reduced mode-locking threshold, which are vital for the additional advancement of mode-locked ultrafast fiber lasers.

Due to the optical nonlinearity of low-dimensional (LD) materials-based SAs, they are able to modulate and control the circulating light wave periodically in the laser cavity, which results in many longitudinal modes to phase oscillation through ultrafast carrier excitation and the recombination process, thus generating regular ultrashort pulse trains in the time scale. Pauli-blocking plays a critical role in an SA, which reduces the light absorption in the SA instantaneously if a large number of electrons are excited from the lower energy level of the SA to the upper energy level by the incidence of larger light fluence [6,20–22]. So far, the emergence of low-dimensional (LD) SA materials, including two-dimensional (2D), one dimensional(1D), and zero-dimensional (0D) materials such as graphene, carbon nanotubes (CNTs), quantum dots, respectively with various advantages over SESAM, provides a new prospect for the development of pulsed fiber lasers because of their distinct structure and physical properties [20–36]. They exhibit divergent physical behaviors varying from semiconductor to insulator and metal to semimetal [37–45]. 2D materials, which are mostly studied and investigated among three types of LD materials, have wide-ranging applications in optics, involving ultrafast fiber lasers as well as modulation, generation, propagation, and detection of light [46–53], have been employed to develop integrated photonic circuits.

2D material belongs to the atomic layer material, which can be a monolayer or multiple-layer thick. 2D materials have solid covalent bonding in a layer and weak interlayer van der Waals force. The electrons' dynamic is limited in the 2D structure if there is no interference of interlayer interaction, which provides 2D materials with numerous novel optical and electrical characteristics [53–55]. Graphene is the earliest monoatomic 2D layer material revealed, with remarkable optical, electrical, mechanical, and thermal behaviors [40–45]. Following 2D graphene, black phosphorus (BP), transition metal dichalcogenides (TMDs), topological insulators (TIs), perovskite and MXene, and other divergent 2D materials were investigated [8,9,56–58]. Moreover, the development of 1D materials such as carbon nanotubes (CNTs) and 0D materials such as quantum dots (QDs) has also made numerous, excellent accomplishments in the development of mode-locked pulsed ultrafast fiber lasers (Figure 1) [22–24,28,59–64]. Since the CNT and graphene were firstly suggested in 2004 and 2009 as an optical SA for mode-locked ultrafast fiber lasers, respectively [21,22], many other LD materials other than graphene, comprising TMDs, TIs, black phosphorus, bismuthine, MXenes, and metal-organic frameworks, quantum dots have been consecutively investigated, implying the significant growth of new SAs based on LD for ultrafast fiber lasers [8,35,58,65]. In addition to these, externally tunable in-line nonlinear LD SAs, where the fiber laser operation can be tuned from continuous wave, through Q-switched to passively mode-locked regime employing electrical gating or external optical bias, have also been demonstrated using various LD SAs [25–27,66–68].

Here, in this review, we study the basic characteristics, renowned synthesis methods of the most widely studied LD materials, and fabrication methods for SA devices comprising various coupling techniques to incorporate LD SAs into fiber systems in brief. Afterward, we mainly focus on the advancement in mode-locked (ML) ultrafast fiber lasers based on various LD SAs (especially 2D and 1D) operating in different dispersion regimes at telecommunication wavelengths near ~1550 nm. In addition, we focus on the recent development of externally controlled fiber lasers based on various LD SAs, which operate in several operating regimes such as QS, QS-ML, and ML depending on the external electrical or optical bias to the SA. In the end, we discuss several prospects about the perspectives and potential advancements of ultrafast pulsed fiber lasers based on LD SA materials.

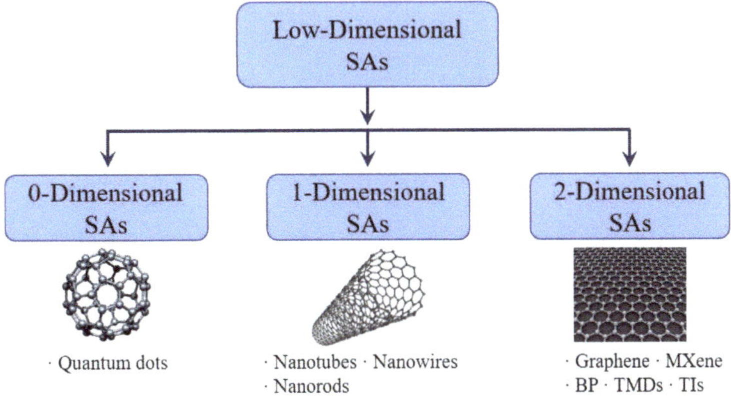

Figure 1. Low-dimensional SAs classification featured as 0D, 1D and 2D structure varieties. [Images are publicly available online].

2. Properties/Characteristics of LD Materials

2.1. 2D Materials

2D materials, in the area of ultrafast nonlinear optics and photonics, are distinguished by their ultrafast recovery, wideband nonlinear saturable absorptions, significant nonlinear refractive indices, and capability as superior mode-lockers for ultrafast fiber lasers [69–76]; what follows is a brief outline of atomic and bandgap structures, and recovery times in 2D materials. Figure 2 illustrates a detailed contrast among the discovered various 2D materials in the family. In this part of the review, fundamental characteristics of related various 2D materials and their incorporation techniques in fiber optic systems are discussed.

2.1.1. Graphene

Graphene, an atomic layered sp^2-bonded carbon atom arranged in a honeycomb lattice [42,45], is considered as the pioneer of all 2D materials available afterward (Figure 2a), expediting huge prospectives for SA in the research field of ultrafast fiber lasers. Monolayer graphene is estimated to absorb a 2.3% incident light infrared (IR) region owing to its gapless Dirac cone structure [40,44]. Unique characteristics of graphene, such as ultrashort recovery time (<200 fs), low saturable absorption (~10 MW/cm^2 [77]), great relative modulation depth (>60% per layer [21]), and wavelength-independent operation (ranging from the visible to the terahertz), make it special and allow it to perform efficiently as an SA to build wideband mode-locked ultrafast fiber laser pulses.

2.1.2. Transition Metal Dichalcogenides (TMDs)

TMDs with the chemical formula MX_2, where M refers to a transition metal (e.g., Mo, Nb, Ti, W) and X refers to a chalcogen (e.g., S, Se, or Te), are a group >40 various semiconductors [78,79]. A TMD monolayer is displayed as a layered structure like graphene, where the single transition metal layer is sandwiched in between two chalcogen layers. TMD exhibits an energy bandgap ranging from 1 to 2.5 eV as a group of semiconductors depending on different chemical compositions (Figure 2b). Surprisingly, incident photon energy on TMDs is considerably lower than their normal bandgaps, undergoing significant absorptions due to the carrier excitation within sub-bandgaps formed by the pristine edge states of TMDs [80,81]. In the meantime, TMDs exhibit ultrashort recovery times in the picosecond scale, useful in the field of ultrafast light modulation and ultrafast fiber lasers. For example, $MoTe_2$, MoS_2, WS_2, and WSe_2 have been extensively studied and investigated to generate ultrafast fiber lasers in 1.5 and 2 μm wavelength regimes.

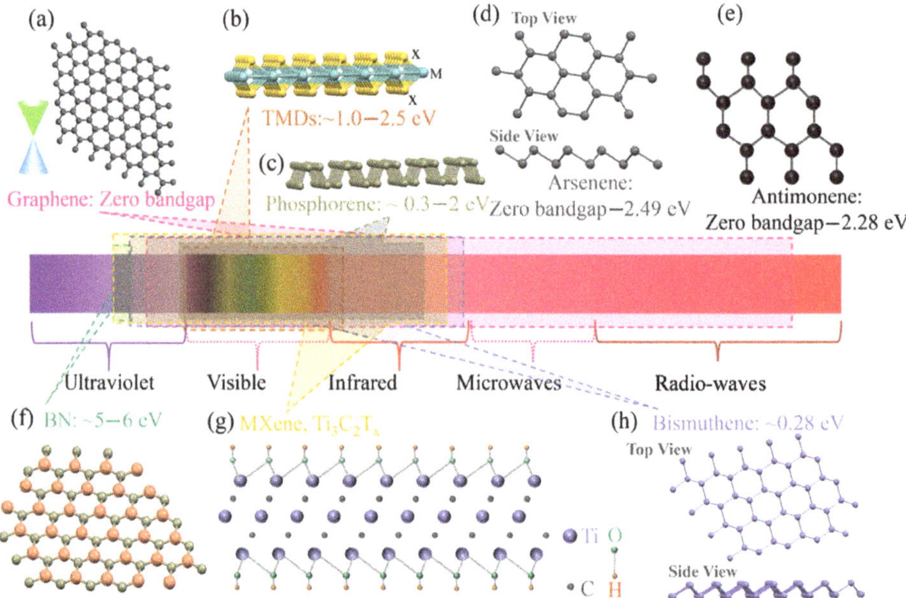

Figure 2. The atomic structure and spectral region of 2D materials. The atomic structure and spectral region of 2D materials such as graphene (**a**), TMDs (**b**), phosphorene (**c**), arsenene (**d**), antimonene (**e**), bismuthine (**f**), MXene (**g**), and BN (**h**). Reproduced with permission [82]. Copyright 2019, Wiley-VCH.

2.1.3. Black Phosphorus (BP)

BP is an allotrope of phosphorus and featured as thermodynamically stable at ambient temperature, also called phosphorene for its monolayer case (Figure 2c) [83,84]. BP is formed as a ring structure linked by six phosphorus atoms as with graphene, where each atom is linked to three neighboring phosphorus atoms. Unlike the graphene structure, the structure of BP is puckered, which breaks the symmetry and results in an angle-dependent optical nonlinearity [85]. BP is featured as a direct bandgap semiconductor that is tunable depending on the number of layers (thickness). The bandgap in BP is ranged from 0.35 eV for a bulk case to 2 eV for monolayer BP, which revealed its wideband nonlinear optical response in the MIR regime [86,87], and extensive investigation for ultrafast fiber lasers [24,88]. It has been exhibited that when a BP nanosheet was excited by various wavelength light with photon energies ranging from 1.55 to 0.61 eV, the recovery time depending on the wavelength was varied from 0.36 to 1.36 ps [89]. At room temperature, BP exposed to air in ambient conditions is unstable and degrades its physical properties, requiring high-quality passivation to improve its stability on a long-term basis [90].

2.1.4. Topological Insulators (TIs)

TIs are identified as novel 2D material along with other 2D materials having topological order protected by nontrivial symmetry, which acts as insulators in their inner portion, but gapless conducting states appear on its surfaces [91–93]. TIs exhibit effective wideband nonlinear optical responses, like graphene, from the visible to the mid-IR owing to their small indirect bulk bandgap of 0.2–0.3 eV. The phonon-induced carriers' lifetime of TIs is as short as several picoseconds, revealing them to be used for ultrafast fiber laser and nonlinear optical modulators. Among the various TIs, Bi_2Te_3, Bi_2Se_3, and Sb_2Te_3 are the most-used TIs as SAs.

2.1.5. MXenes

MXenes categorized as a new class of 2D transition metal nitrides, carbonitrides, or carbides, with the composition of $M_{n1}X_nT_xn$, where M refers to transition metals (Ti, Sc, Hf, Zr, Nb, V, Cr, Mo, Ta, etc.), X stands for nitride and/or carbide, and T_x is surface terminations (O, OH, F, etc.) [94,95]. Few-layer $Ti_3C_2T_x$ exhibits an indirect energy bandgap lower than 0.2 eV and reduced absorption of around 1%/nm (Figure 2g). 2D MXene materials are usually periodically stacked by means of van der Waals interaction in which no internal termination occurs, a similar feature seen in phosphorene, graphene, and TMDs. In the latest work, it was observed that stacked MXene comprises a set of monolayer MXene in it without any significant disorder [33], implying the possibility of using MXenes as an SA to generate ultrafast fiber lasers, which avoids the deleterious methods of monolayer dispersion.

2.1.6. Bismuthine

Bismuthine has drawn immense attention in the scientific community, owing to its unique electronic and mechanical characteristics, along with its excellent stability [37,96]. In the latest report, the tunable optical bandgap depending on the layer number varied from nearly 0 to 0.55 eV (Figure 2h) in beta-bismuthine, which reveals bismuthine as an effective wideband nonlinear optical material from the terahertz regions to the near-IR [97]. Comparably shorter recovery time of 2.8 ps in monolayer bismuthine also implies that bismuthine can be a potential candidate as an SA for ultrafast fiber laser applications.

2.1.7. Other 2D Materials

The 2D materials mentioned above display distinctive yet complementary characteristics and, therefore, new opportunities for optical applications in ultrafast fiber lasers [98–100]. However, the SA technology required for standard fiber laser operation is always anticipated as perfect as it is theoretically attributed with superior optical properties including ultrashort carrier lifetimes, elevated modulation depths, and damage thresholds. Therefore, it is obvious that exploring novel 2D materials as SA is continuous since it was first implemented. In addition, the controllability of key features in current 2D SAs and engineering in the laser cavity with tunable behavior of SAs also offers more functionalities in the development of controllable ultrafast fiber lasers [25–27,66,68,101]. Combining two or more similar or different 2D materials and building van der Waals heterostructures prospects towards developing multi-functional, high efficiency, broadband controllable photonic devices [102,103]. Recently, several ultrafast fiber lasers mode-locked by heterostructure SAs have been reported [104,105].

2.2. 1D Materials

Carbon Nanotubes (CNTs)

Carbon nanotubes, a unified cylindrical, 1D nanocrystalline graphite material with a high aspect ratio, include a diameter varying from a few to several hundred nanometers and length of up to centimeters. According to the number of tube walls, CNTs can be divided into many types: single-walled CNTs (SWCNTs), double-walled CNTs, and multi-walled CNTs. The chiral properties of carbon tubes lead to different applications of metallic and semiconducting carbon tubes. The carbon tube of the semiconductor type has an obvious bandgap, while the band of the metal type is continuous. CNTs possess multiple excellent properties and advantages that are well fitted with the requirements of a good SA. The measured third-order nonlinear polarizability by pump-probe spectroscopy is 10^{-7}–10^{-10} esu (1 esu = 1.11×10^{-9} m^2 V^{-2}). The recovery time was measured to be composed of a fast intraband carrier relaxation time of 0.3–1.2 ps and a slow recombination process of 5–20 ps [106]. Moreover, the superior thermal conductivities as high as 5000 W m^{-1} guarantee intrinsic high-power handling. A highly developed growth process significantly lessens the price of raw materials as well as the research cost. More notably, the expansion of CNT SAs over the past 15 years simplified the direction of all-fiber integration configuration, and extensive investigations have validated its operation in a broadband range, which is an underlying disadvantage of industrial SESAMs. Therefore, CNT SAs are a trustworthy candidate to perform as a promising replacement to SESAMs in the future.

3. Synthesis of LD SA and Device Fabrication

3.1. Synthesis Techniques

In the past decade, several synthesis techniques have been successfully established to prepare the LD materials introduced in the previous section. The most common synthesis methods for LD materials are categorized into two types: top-down and bottom-up methods. To briefly review, a few standard methods from these types are discussed. The top-down exfoliation methods comprise liquid-phase exfoliation (LPE), mechanical exfoliation (ME), laser thinning, and chemical exfoliation, where single-layer or multiple-layer 2D nanosheets are separated from bulk materials by violating the van der Waals force between layers [9]. Bottom-up methods include pulsed laser deposition (PLD) and chemical vapor deposition (CVD), where high-quality 2D materials in atomic layer scale are effectively synthesized by explicitly adjusting the chemical reactions among solid precursors. Most common and widely applied synthesis and preparation methods for 2D SAs to realize mode-locked fiber lasers will be discussed in brief.

3.1.1. Mechanical Exfoliation (ME)

The ME technique is commonly used in the manufacturing of atomically and few layers thick sheets of 2D layered inorganic materials [44,107–110]. Researchers can acquire high-quality 2D mono- and few-layer materials by resolving the van der Waals force and splitting layers away from bulk materials. This method was first used in the discovery of 2D graphene from graphite flakes in 2004 by Geim and Novoselov [44] owing to its flexibility and potential to manufacture few-layer materials with outstanding qualities. Compared to the bulk materials, single-layered or multi-layered 2D materials are highly comprehensive and have negligible defects, making them ideal for basic scientific study. Although this method is convenient, fast, and cost-effective, it does have some drawbacks. Since large-area single-layer 2D materials synthesis using this method is difficult, it is only appropriate for fundamental research in a laboratory. Many studies have demonstrated ME using scotch tape to synthesize other 2D materials. This procedure is often used to obtain monolayer BP. Zenghui Wang et al. employed key strategies specifically designed to expedite the transfer of BP after exfoliation to reduce the material's exposure to the ambient conditions [111].

3.1.2. Liquid Phase Exfoliation (LPE)

Liquid phase exfoliation with a high yield has become a viable alternative to mechanical exfoliation [50,112], where large numbers of dispersed 2D layers are exfoliated from its bulk (layered compound) state in liquids. There are four LPE approaches mostly used to eliminating interlayer forces: (i) oxidation followed by ultrasonication, (ii) ion intercalation, (iii) ion exchange, and (iv) sonication-assisted exfoliation. In the oxidation method, layered 2D crystals with low reductive potential are exfoliated by oxidation and subsequent dispersion and ultrasonication in suitable solvents. In ion intercalation, the interlayer gap is expanded by embedding organic or ionic materials as intercalants, such as *n*-butyllithium or IBr in liquids, which disrupt the interlayer adhesion force between the 2D layers in bulk. In ion-exchange methods, layered compounds contain ions between the layers so as to balance the surface charge on the layers. These ions are exchanged in a liquid environment for other larger ions, leading to substantial swelling in layered compounds, and subsequent agitation results in an exfoliated dispersion. The other method is to create microbubbles and pressures between layers of bulk materials directly using high-intensity ultrasonics. However, these approaches can effectively modify the material's composition in liquids, and the quality must be enhanced further.

3.1.3. Chemical Vapor Deposition (CVD)

The uncontrollable scale and random thickness of few-layered 2D materials acquired using the ME or LPE techniques are counterproductive to the efficiency of an SA [113,114]. CVD is an essential bottom-up approach for synthesizing comparatively large area 2D materials with scalability. In 2009, the CVD technique was first employed to synthesize graphene on a copper substrate with a large area (centimeter scale) and high uniformity [115]. Following this synthesis, numerous 2D materials including TIs and TMDs were synthesized employing the CVD method. Few-layer WSe_2 with a large area and improved quality were synthesized by CVD to be employed as SA in the generation of ultrafast fiber laser, reported by W. Liu et al. as shown in Figure 3 [116]. One of the key features of the CVD technique is its functionality to adjust the number of layers in 2D materials, which elevates the modulation depth of 2D SAs. Nevertheless, this method undergoes deleterious transfer steps from the grown substrate to the fiber system to build the SA device, making this technique complicated to realize cost-effective devices [21,116,117]. To solve these issues, direct synthesis methods have been reported for graphene to build nonlinear optical devices including SA devices [32,101,118–120].

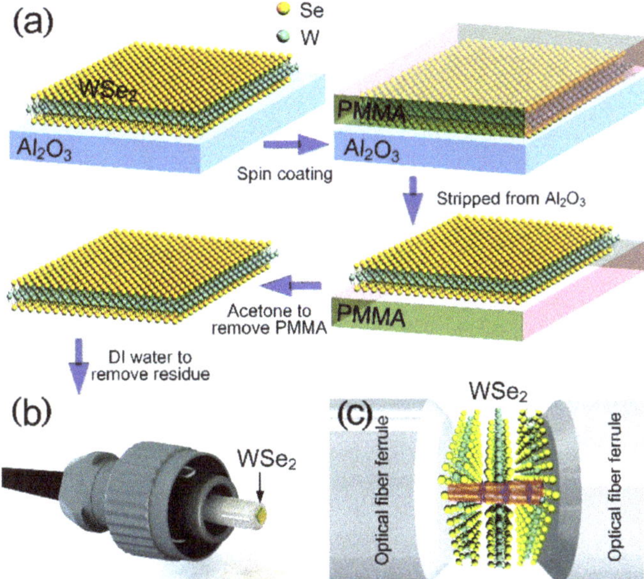

Figure 3. Graphic illustration of the preparation of the WSe_2-based SA by the CVD method. (**a**) The transfer scheme of WSe_2 films. (**b**) optical image of WSe_2 films transferred onto the fiber ferrule end facet. (**c**) schematic presentation of the few-layer WSe_2 and light interaction. Reproduced with permission [116]. Copyright 2018, RSC.

3.2. LD SAs Integration with Optical Fiber

LD materials should be transferred, synthesized, or directly deposited onto optical fibers to build in-line SA devices. Additionally, a significant interaction of LD material with guided light is required for the stable operation of all-fiber mode-locked ultrafast fiber lasers. Usually, these coupling techniques are variously beneficial for different fiber laser schemes. For LD SA-based fiber lasers, the incorporation of SA must be done with the optical fiber or its component. Figure 4 illustrates a variety of mostly utilized fiber coupling methods which are classified into two cases: transmission coupling and evanescent-field coupling.

3.2.1. Direct Coupling

Direct coupling is the most widespread method to realize SA devices, where SA is sandwiched directly between two fiber end facets, as illustrated in Figure 4a. Various LD materials including mechanically exfoliated graphene, CVD-grown graphene or TMDs, 1D CNT, and MBE-grown TIs have been incorporated in fiber end facets using this method in the previous research [121]. Direct coupling facilitates the stable interaction between the guided laser signal and the SA, resulting in reliable mode-locking operation. Nevertheless, if the incident laser power is relatively high, the direct interaction scheme often leads to significant damage to the SA. Thus, a direct coupling scheme with sandwiched SA materials is generally suitable for mode-locked ultrafast fiber lasers operating at low average power.

3.2.2. Evanescent-Field Coupling

To reduce the damage threshold of fiber SAs, a widely used alternative technique called evanescent-field coupling was introduced to incorporate LD SAs onto a tapered fiber and a D-shaped fiber (also called side-polished fiber, SPF) as shown in Figure 4c,d. In 2007, Yamashita et al. for the first time introduced evanescent field coupling of CNTs on SPF and tapered fiber [122,123]. The evanescent field of guided light through the fiber in this scheme interacts with LD SA on the fiber surface. As shown in Figure 4b, SA materials can be filled into a hollow photonic crystal fiber (PCF) and a hollow core fiber (HCF), which are linked to the fiber laser cavity [124,125]. To achieve this, LD SA materials are initially dispersed in a solvent followed by the subsequent filling of SA dispersion into the hollow PCF and HCF. Then the SA-filled PCF or HCF is dried and connected to the fiber laser ring cavity [125]. However, the core size of PCFs being in the μm range, it is challenging to dry the solvent completely inside the core, resulting in elevated insertion losses, hence the unstable performance of ultrafast fiber lasers.

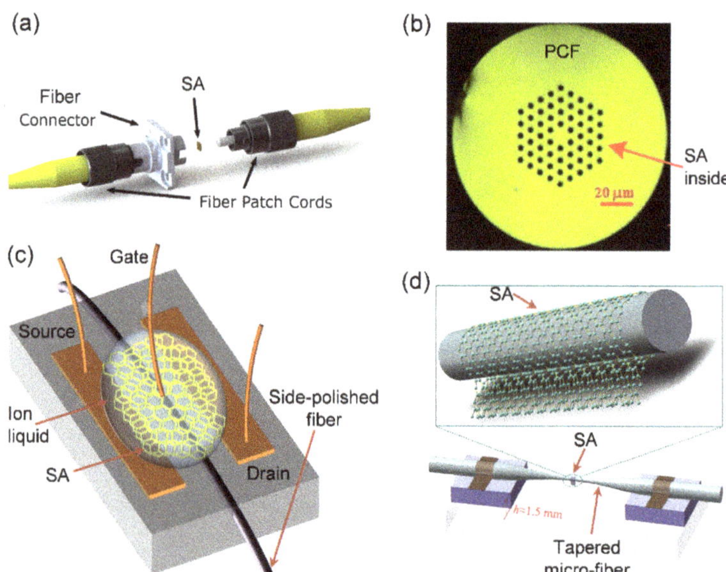

Figure 4. Various schemes for incorporating LD SAs in optical fiber. (**a**) SA sandwiched between two fiber connectors. Reproduced with permission [126]. Copyright 2015, The Optical Society of America. (**b**) LD SA injected inside a hollow photonic crystal fiber (PCFs). Reproduced with permission [125]. Copyright 2013, The Optical Society of America. (**c**) LD SAs transferred onto SPF surface. (**d**) depositing SAs around tapered micro-fiber. Reproduced with permission [127]. Copyright 2016, Springer Nature.

After incorporation of LD SA in the fiber system, the SA module must be examined to ensure the successful incorporation was occurred or not. The most common technique to characterize the SA module is nonlinear transmission to analyze the modulation depth, saturation fluence, and non-saturable loss of the SAs in a fiber setup [20,21,25].

4. Ultrafast Fiber Laser Based on LD SA

In the field of SA-based fiber lasers, mode-locking and Q-switching are two key techniques to generate ultrashort pulses. Both the techniques are useful and selective for various applications depending on their different advantages. Q-switched lasers are mostly utilized in laser processing and military purposes owing to their high pulse energy. On the other hand, because of the ultrashort pulse duration in the femtosecond scale, mode-locked lasers are employed in the areas of nonlinear imaging, micromachining, and fundamental scientific research. Figure 5 shows a comparatively compact and convenient all-fiber laser ring cavity where the optical isolator, output coupler, and wavelength division multiplexer are combined in a hybrid component to reduce the size of the cavity [128]. Over the last decade, numerous LD SAs have been introduced following the invention of graphene and CNT as SA to build ultrafast fiber lasers using the typical fiber laser cavity mentioned above. A summary of various LD SA-based mode-locked Erbium-doped ultrafast fiber lasers' basic characteristics is provided in Table 1. In terms of the performance of these mode-locked lasers based on graphene, some exciting results have been obtained, including the minimum pulse width and maximum output power of 29 fs [129] and 174 mW [130], respectively. For TIs, the corresponding figures were 128 fs [131] and 75 mW [132], for TMDs they were 67 fs [133] and 57 mW [134], and for BP they were 102 fs [135] and 5.6 mW [114], respectively. Below we introduce the mode-locked ultrafast fiber laser based on several LD SAs and review the advancement of MLFL in terms of controllability in operation, performance, and application.

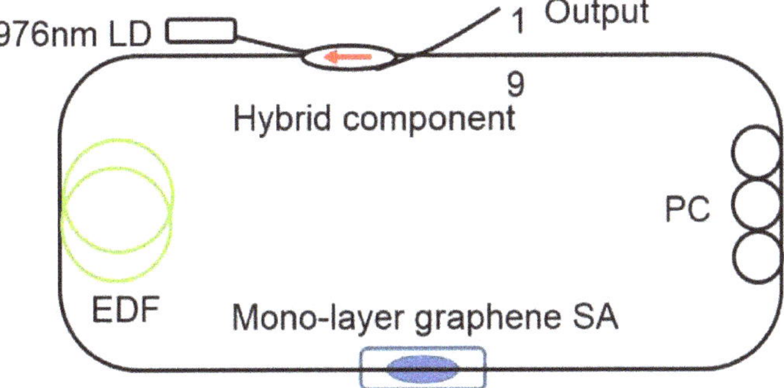

Figure 5. Schematic illustration Er-doped fiber laser ring-cavity comprising a single layer graphene SA on the SPF. LD: laser diode; EDF: erbium-doped fiber; PC: polarization controller; hybrid component: an integrated wavelength-division multiplexer and isolator. Reproduced with permission [128]. Copyright 2015, The Optical Society of America.

4.1. Ultrafast Fiber Laser Based on Graphene SA

After the discovery of graphene, it has been widely studied and investigated in various scientific research fields because of its distinctive nonlinear optical properties [31,54,136]. The features of broadband optical absorption, ultrafast carrier recovery, gapless band structure, and high damage threshold make graphene one of the leading SAs in the application of fiber laser lasers. First, the convergence of valence band and conduction band at the Dirac point of graphene denotes a gapless semi-metallic band, i.e., zero-bandgap structure, which results in graphene being conducive to wideband absorption. Secondly, the presence of an exclusive quantum tunneling effect in graphene carriers implies fast relaxation time and higher carrier. The relaxation time was investigated through pump-probe experiments in graphene by Bao et al., and the fast relaxation time of 150 fs has been measured in the case of graphene [137]. They exhibited that, to shape ultrashort laser pulses in fs scale, the graphene with a fast relaxation time is more effective. Finally, graphene has a melting point of up to 4510 K [138]. In the experiment, a high damage threshold with graphene SA indicates an appropriate application of graphene SA in high-power lasers. To date, numerous studies and investigations with graphene in ultrafast fiber lasers have been performed. Bao et al. designed and created graphene SA devices for the first time. They investigated graphene SAs as nonlinear optical materials which were employed to generate stable mode-locked ultrafast Erbium-doped fiber laser (EDFL) with 3 dB bandwidth of 5 nm operating at 1565 nm. Following this work, graphene SA-based MLFL has been a hot topic to investigate the various schemes and techniques for controllable MLFL based on graphene SA. In addition, Park et el. utilized the evanescent field coupling of monolayer graphene to produce graphene SA on SPF [128]. Through numerical study, they showed that a strong optical absorption of more than 90% can be achieved without significant scattering loss in a monolayer-graphene saturable absorber by employing an index-matched over-cladding structure on the graphene/SPF. By tuning the index of over-cladding, they effectively controlled and tuned the light coupling into the laser cavity, which results in significant control of the MLFL characteristics. Figure 6 shows the optical spectrum of the laser output, where the measured average output power was 5.41 mW at an applied pump power of 120 mW. The typical spectral shape of an optical soliton was observed, where the measured spectral bandwidth was 8.6 nm at 1607.7 nm. Assuming a soliton pulse, the pulse duration measured by an intensity autocorrelator was 377 fs, as shown in Figure 6b, the fundamental repetition rate was measured to be 37.72 MHz (Figure 6c). The pulse width of the mode-locked laser reduced with the over-cladding index increased, where the measured pulse widths were 429, 395, and 377 fs for the over-cladding indices of 1.426, 1.434, and 1.444, respectively (Figure 6d).

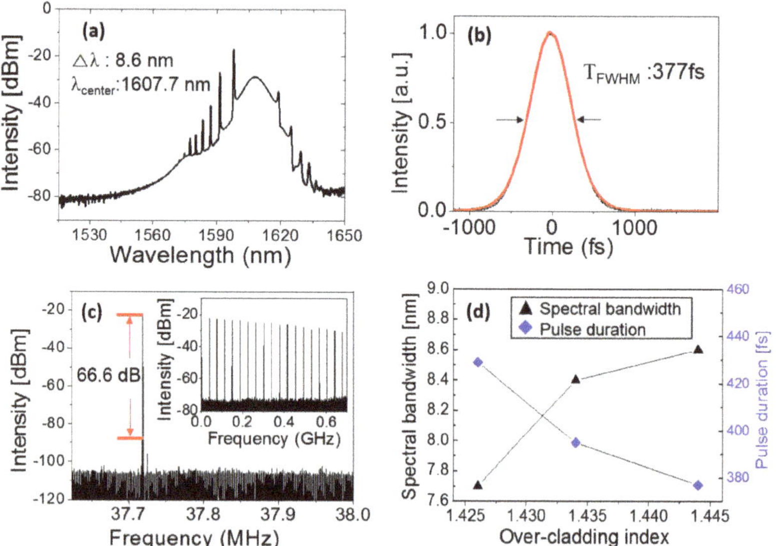

Figure 6. Mode-locked ultrafast fiber laser based on graphene SA (**a**) soliton pulse spectrum and (**b**) autocorrelation trace of the mode-locked pulse. (**c**) RF spectrum of the laser output pulse train (inset: RF spectrum viewed over a wide frequency range) and (**d**) 3-dB bandwidth and pulse width of the implemented laser as functions of the over-cladding index of the SA. Reproduced with permission [128]. Copyright 2015, The Optical Society of America.

Table 1. Summary of the characteristics of various LD SA-based ultrafast fiber lasers operating at ~1.55 μm.

SA	Center Wavelength (nm)	3 dB Bandwidth (nm)	Pulse Duration (ps)	Repetition Rate (MHz)	Output Power (mW)	Pulse Energy (nJ)	Ref.
Graphene (2D)	1559.12	6.16	0.432	25.67	-	0.09	[139]
	1566	4.92	0.88	6.22	-	-	[140]
	1555	6	0.59	45.88	0.91	-	[141]
	1545	48	0.088	21.15	1.5	0.071	[142]
	1553	3	1	8	1	0.125	[143]
	(CS)1565	7	13.8	25.8	0.7	10.2	[130]
	(DS)1559	10.4		16.99	174		
	1607.7	7.7–8.6	0.37~0.429	37.72	5.41	-	[128]
CNT (1D)	1547.5	0.3	22.73	10.61	11.21	1.057	[144]
	1560.1	4.3	0.763	62.2	0.445	0.007	[145]
	1555.1	3.9	0.85	10.89	3.19	0.29	[146]
	1564.5	5	0.57	18.3	0.316	0.017 pJ	[147]
	1563	12.1	12.7	9.8	335	34	[60]
	1560	4.83	0.602	11.25	8.58	0.763	[64]
	1560	4.33	22.2	0.51	4	0.18	[61]
	1560	42	0.093	38.117	11.2	0.3	[148]
BP (2D)	1569.24	9.35	0.280	60.5	-	-	[149]
	1566.5	3.39	0.94	4.96	5.6	-	[114]
	1561.1	3.25	0.8	5.86	0.3	0.051	[150]
	1555	40	0.102	23.9	1.7	0.071	[135]
Bi$_2$Se$_3$ (TI)	1557.5	4.3	0.66	12.5	1.8	0.144	[30]
	1558.3		3.01	5.1			
	1557.4	0.9	3.42	(HML)388	-	-	[151]
	1559.4		2.02	(HML)239			
	1600	7.9	0.36	35.45	0.86	-	[152]
	1554.56	7.91	0.908	(CS)20.27	5.5	0.27	[132]
	1559	26	7.564	(DS)7.04	75	0.27	
	1562.4	4.28	0.630	23.3	-	0.0156	[153]
	1557.908	0.342	7.78 ns	1.71	82.6	48.3	[154]
Bi$_2$Te$_3$ (TI)	1570	5.88~6.66	0.403	28.5	-	-	[155]
	(DS)1560	5.6	2.7 ns	1.7	32.9	19.3	[156]
	(CS)1558.5	0.95	1.22	4.88	5	1.02	[157]
	(HML)1558.5	1.08	2.49	2.04 GHz	5.02	-	
	1547	4.63	0.6	15.11	0.8	0.0529	[107]
	1558.459	1.696	3.22 ns	1.704	40.37	23.9	[158]
	1560.8	9.15	0.286	18.55	0.5	0.027	[159]
MoS$_2$ (TMD)	1570.1	2.7	1.36	5.924	3.5	0.59	[29]
	1571.8	3.5	0.83	11.93	5.85	0.49	[160]
	1574.6	9.5	0.79	29.5	4.13	0.14	[161]
	1568	23.2	4.98	(DS)26.02	-	-	[162]
	1568	12.38	0.637	(CS)33.48			
	1556.86	2.47	-	6.77	0.065	0.01	[163]
	1560	20.5	0.2	14.53	1	0.069	[113]
	1569.5	4	0.71	12.09	-	-	[164]
	1556.3	6.1	0.935	463	5.9	-	[165]
	(CS) 1530.4	2.1	1.21	8.968	-	-	[166]
	(BS)	(period) 2	1.2	8.968	-	-	
MoSe$_2$ (TMD)	(CS)1557.3	5.4	0.798	15.38	-	-	[167]
	(HML)1557.3	5.1	0.751	3.27 GHz	0.23~22.8	14.6~6.7 pJ	
	1560	7.8	0.580	8.8	-	0.0913	[168]
	1552	12.72	0.207	64.56	-	-	[169]
	1558.35	2.9	1	16.27	-	-	[170]
MoTe$_2$ (TMD)	1561	24.9	0.1119	96.323	23.4	-	[171]
	1532.5	1.5	2.57	6.95	1.7	-	[172]
	1559.57	11.76	0.229	26.601	57	2.14	[134]

Table 1. *Cont.*

SA	Center Wavelength (nm)	3 dB Bandwidth (nm)	Pulse Duration (ps)	Repetition Rate (MHz)	Output Power (mW)	Pulse Energy (nJ)	Ref.
WS$_2$ (TMD)	1565	8.23	0.332	31.11	0.43	-	[173]
	1566	5.6	0.457	21.07	0.32	-	[173]
	1540	114	0.067	135	-	-	[133]
	1572	5.2	0.595	25.25	-	-	[174]
	(DS)1565.5	14.5	21.1	8.05	1.8	0.22	[175]
	1558.5	-	0.675	19.58	0.625	-	[176]
	1563.8	5.19	0.524	19.57	2.64	0.134	[177]
WSe$_2$ (TMD)	1556.42	6.06	0.477	14.02	-	-	[117]
WTe$_2$ (TMD)	1556.2	4.14	0.77	13.98	-	-	[178]
MXene	1550	42.54	0.104	20.03	-	0.065	[179]
	1567.3	3.1	0.946	8.24	-	-	[56]
	1557	5	0.66	15.4	0.05	-	[33]
	1555.01	22.2	0.159	7.28	3	0.41	[180]
	1565.4	3.4	5.3	8.25	-	-	[181]
BP QD (0D)	1561.7	3	0.882	5.47	-	-	[24]
	1560.3	2.2	1.2	5.62	2.23	0.45	[182]

The corresponding spectral bandwidth changed from 7.7 nm to 8.4 nm to 8.6 nm, indicating that the SA possesses different pulsating abilities depending on the over-cladding index, which affects the pulse formation for a given laser cavity. Consequently, this may allow one to control the soliton pulse bandwidth and pulse width of the generated ultrafast laser through appropriate selection of the over-cladding material with various refractive indexes.

Although ultrafast fiber lasers based on graphene SA have been developed extensively, numerous deficiencies with graphene SA have progressively been revealed through growing investigation. Firstly, the absorption efficiency of 2.3% per layer results in reduced modulation depth which limits its advanced application in ultrafast fiber lasers. Even if a higher number of graphene layers elevate the modulation depth, the unsaturated loss, in this case, declines the performance of the fiber laser. Secondly, the tunable operating wavelength plays a critical role in the saturation threshold of graphene SA. A shorter operating wavelength exhibits a higher saturation threshold in graphene SA. This implies graphene as comparatively proper SA for fiber laser operating in the mid-infrared range but marginally worse performance while working in shorter wavelengths.

4.2. Ultrafast Fiber Laser Based on CNT SA

The first verification of CNTs as SA in an ultrafast fiber laser system took place in 2003 by Set et al., and ultrashort pulses of ~1 ps were demonstrated in 1550 nm [183]. Subsequently, CNT SAs have been rapidly adopted by many research groups. The most popular type of SWNT SA used in fiber lasers concentrates on polymer composite film, which is mechanically or optically sandwiched between fiber connectors. Another approach based on the evanescent field can be a candidate, and SWNTs can be coated on a D-shaped fiber [122], microfiber [184], or injected into micro slots [185,186], photonic crystal fiber [187], and hollow-core optical fiber (HOF) [188]. Choi et al. demonstrated the nonlinear interaction scheme of SWCNTs with laser light by using the HOF, which offers the advantages of robust, efficient, and long interaction of guided light with SWCNTs with a simple fabrication process (inset of Figure 7a) [188]. The HOF filled with low concentration SWCNT/polymer composite exhibiting broadband absorption is prepared as an in-line SA, resulting in a passively mode-locked fiber laser with the spectral bandwidth; the pulse duration and repetition rate of the laser output are 5.5 nm, 490 fs, and 18.5 MHz, respectively (Figure 7b,c) [188]. In another demonstration, a dissipative soliton fiber laser with high pulse energy (>30 nJ) based on a single-walled carbon nanotube saturable absorber (SWCNT-SA) on SPF was reported by Jeong et al. [60].

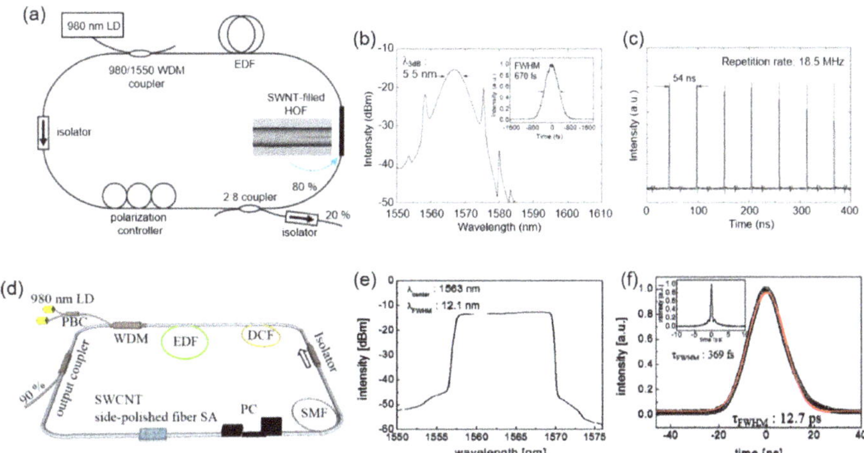

Figure 7. 1D SWCNT SA-based mode-locked fiber laser: (**a–c**) Conventional soliton: (**a**) Schematic of the fs fiber laser using the SWCNT-filled HOF. The inset figure shows the spliced image between the normal SMF and HOF, where adiabatic mode transition occurs. (**b**) Measured optical spectrum and pulse duration (inset) of the mode-locked fiber laser. (**a–c**) Reproduced with permission [188]. Copyright 2009, The Optical Society of America. (**c**) The output pulse train of the laser shows a repetition rate of 18.5 MHz. (**d–f**) Dissipative soliton: (**d**) Configuration of the fiber ring laser including the SWCNT-SA and the DCF. (**e**) The optical spectrum of the mode-locked laser at net cavity dispersion of 0.087 ps^2 and (**f**) Measured pulsed duration fitted with Gaussian pulse. The inset shows the pulse compressed by additional SMF at extra-cavity. (**d–f**) Reproduced with permission [60]. Copyright 2014, The Optical Society of America.

A laser cavity generates a dissipative soliton mode-locked pulsed laser only when the laser cavity net dispersion is in the normal dispersion regime. A dispersion compensating fiber (DCF) with large normal dispersion at 1550 nm was inserted along with SMF-28e (anomalous dispersion at 1550 nm) in the cavity to realize the normal net cavity dispersion, as shown in Figure 7d. Stable passive mode-locking of a dissipative soliton laser was obtained at a net cavity dispersion of around 0.141 ps^2. The generated dissipative soliton exhibits spectral bandwidth of 12.1 nm with a flat top spectral shape at the central wavelength of 1563 nm, as shown in Figure 7e. The flat-top behavior of the generated soliton pulse is evidence of dissipative soliton mode-locked pulsed laser. The laser stably delivers linearly chirped pulses with a pulse duration of 12.7 ps (Figure 7f), and the average power of the laser output is measured as 335 mW at an applied pump power of 1.27 W. The corresponding pulse energy is estimated to be 34 nJ at the fundamental repetition rate of 9.80 MHz; this was the highest value reported in all-fiber Er-doped mode-locked laser using an SWCNT-SA.

4.3. Ultrafast Fiber Laser Based on Other 2D SAs

Following an extensive investigation with 2D graphene SA and 1D CNT SA for ultrafast fiber laser, other LD materials-based SAs have also attracted attention to build ultrafast mode-locked fiber laser. In 2014, black phosphorus started to achieve prevalent curiosity due to its unique electro-optical properties [84]. The interesting feature of BP is its band structure which can vary depending on the thickness or number of BP layers. The bandgap of BP reduces with increasing the number of layers or BP thickness because of interaction among the BP layers. The bandgap of bulk BP is 0.3 eV, which estimates that BP is a highly suitable contender for both mid-infrared and near-infrared SAs. One of the unique features of BP is its direct bandgap regardless of the changes in thickness. Compared to graphene, the relaxation time in BP has been found to be faster in the mid-infrared and near-infrared region revealed by Wang et al. using a pump-probe experiment, which suggested the higher ability for pulse narrowing compared to graphene [89]. Consequently, BP, with its direct bandgap feature, provides valuable aspects towards the application of ultrafast nonlinear optics and optoelectronics. Numerous ML ultrafast fiber lasers based on BPs have been successfully investigated and recognized. Chen et al., for the first time, successfully constructed a BP-based SA and engaged it in a fiber laser ring cavity to generate stable ultrafast mode-locked fiber laser operating at 1571.45 nm with a pulse width of 946 fs and 3 dB bandwidth of 2.9 nm [110]. The stability of the laser was confirmed by a signal-to-noise (SNR) ratio of 70 dB. Along with the EDFL operating at 1.55 µm, BP-based SA has also been examined in the 1 µm region achieved by YDFL. Hisyam et al. demonstrated MLFL using BP SA in a YDFL ring cavity, which generates 1 µm mode-locked soliton pulse with 7.54 ps and the highest ever output power of 80 mW at their time of report [189]. The pulse energy was measured to be 5.93 nJ. Pawliszewska et al. obtained 2 µm holmium-doped all-fiber lasers having a pulse duration of 1.3 ps, centered at 2094 nm with a bandwidth of 4.2 nm using BP SA [190]. Significant work on BP SA in the divergent wavelength range of 1–2 µm reveals the broadband absorption characteristics of BP. Jin et al. achieved the shortest pulse duration using BP-based SA at the time of their report [135]. A highly functional inkjet printing technology with high scalability was employed to fabricate the SA based on BP. The exfoliated BP flakes are shown in Figure 8a. Ultrafast fiber laser with this BP SA in an EDFL ring cavity exhibited the pulse width of 102 fs, 3 dB bandwidth of 40 nm at 1555 nm, and as shown in Figure 8a–c. Although the ultrafast fiber laser based on BP SA has been developed extensively with its wideband absorption characteristics and the ability to shorten the pulse width, numerous deficiencies with BP SA have progressively been revealed through growing investigation. Mainly, its instability to environmental factors such as humidity and temperature is detrimental to the fiber laser system. The physical properties of BP are highly sensitive to air. BP increases in its volume unexpectedly if it is exposed to air due to its large affinity for water.

Consequently, the BP surface decayed as time goes by if it was not efficiently passivated for a long time [192]. The optical properties of BP are certainly affected by its environmentally unstable behavior, thus influencing the execution of high-performance fiber lasers with BP SA. Moreover, the BP SA is prone to damage under high power laser due to its unavoidable thermal effects in the air, further limiting its application in elevated power regimes [114].

Inspired by the domination of 2D graphene as an SA for fiber lasers, few-layer 2D bismuthine also has been introduced by Lu et al. as an SA, owing to its direct bandgap at 1550 nm. Bismuthine's optical bandgap is tunable and controlled by changing the number of layers. By increasing the layer number from one to six layers, it exhibits an optical bandgap varying from 1.028 eV to 0.747 eV [191]. They investigated few-layer bismuthine using various tools including atomic force microscopy (AFM), which confirmed the thickness of prepared few-layer bismuthine SA as 4 nm with a smooth surface as shown in Figure 8d. Generated ultrafast mode-locked fiber laser in the anomalous dispersion regime operating at 1559.18 nm has a spectral bandwidth of 4.64 nm and pulse duration of 652 fs (Figure 8e,f)

MXene, as a recently developed new 2D material, has attracted considerable attention because of its graphene-like but highly tunable and tailorable electronic/optical properties. Researchers found that a typical MXene has efficient SA with negligible lossy nonlinear absorption components in the spectral range 800–1800 nm, which is indicated in a typical MXene, $Ti_3C_2T_x$, which was deliberately chosen as the investigation object to highlight broadband nonlinear optical response in the near-infrared region [180]. In 2018, Jiang et al. investigated the broadband non-linear photonics of $Ti_3C_2T_x$ by depositing $Ti_3C_2T_x$ solution onto a side-polished fiber [180], where stable mode-locking was achieved in both Yb- and Er-doped fiber lasers operating at the wavelength of ~1 µm and ~1.55 µm, respectively.

Er-doped fiber laser results are depicted in Figure 8g–i. Highly stable self-started CWML is readily obtained when the pump power is above 60 mW. The ultrashort pulse duration of 159 fs was obtained with the spectral bandwidth of 22.2 nm operating at 1555.05 nm. The output power, pulse energy, and peak power at the pump power of 238 mW are 3 mW, 410 pJ, and 2578.6 W, respectively.

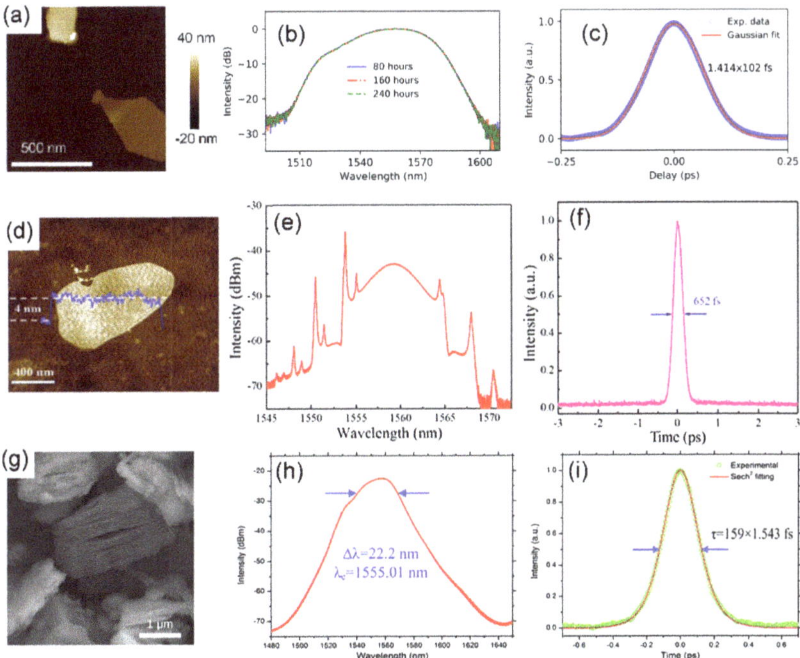

Figure 8. Mode-locked ultrafast fiber laser in the NIR region with various 2D SAs such as BP (**a–c**), Bismuthine (**d–f**), and $Ti_3C_2T_x$ MXene (**g–i**). (**a**) AFM micrograph of selected exfoliated BP fakes; (**b**) Optical spectrum with a bandwidth of 40 nm acquired after 80 h (blue curve), 160 h (red dot line), and 240 h (green dot line), respectively; (**c**) Autocorrelation trace with a Gaussian fit. (**a–c**) Reproduced with permission. Copyright 2018, The Optical Society of America [135]. (**d**) AFM image of few-layer bismuthine. (**e**) mode-locked soliton pulse spectrum with 3 dB bandwidth of 4.64 nm, and (**f**) autocorrelation trace of mode-locked ultrafast laser showing the pulse width of 652 fs. (**d–f**) Reproduced with permission. Copyright 2018, Wiley-VCH [191]. (**g**) SEM image of $Ti_3C_2T_x$, (**h**) mode-locked soliton pulse spectrum with 3 dB bandwidth of 22.2 nm, and (**i**) autocorrelation trace of mode-locked fiber laser showing the pulse width of 159 fs. (**g–i**) Reproduced with permission. Copyright 2017, Wiley-VCH [180].

4.4. Ultrafast Fiber Laser Based on TMD and TI SAs

TMDs, because of their various types with numerous members in the group, have inhabited the effective prominence as prospective contenders of SAs [162,175,178,193–197]. Among other candidates after graphene, which has been extensively researched in the non-linear optics field, TMD is discovered to execute well in terms of ultrafast carrier dynamics, switchable bandgap, and higher-order nonlinear optical response. At present, some TMD materials such as MoS_2, $MoTe_2$, $MoSe_2$, WSe_2, and WS_2 have produced crucial break-throughs in ultrafast fiber lasers [162,175,178,193–197]. For example, the MoS_2 bandgap transforms from indirect to direct while the thickness decreases from bulk to monolayer, along with the bandgap rising from 1.8 to 1.29 eV [198]. Moreover, MoS_2 exhibits a larger third-order nonlinear optical response compared to graphene. In addition to this, MoS_2

highlights a carrier lifetime of nearly 100 ps and ultrafast intraband relaxation time as short as 30 fs [196]. Table 1 includes the properties of ultrafast fiber lasers with various TMD SAs. Among these TMDs, layered MoS_2 was first studied. In 2014, the SA behavior of few-layer MoS_2 was initially observed [194,197,199]. Both the conventional soliton and dissipative soliton mode-locked fiber laser have been reported by individual groups using anomalous dispersion and normal dispersion fiber laser ring cavity, respectively. Xia et al. demonstrated an Er-doped ultrafast fiber laser passively mode-locked by a multilayer MoS_2 SA prepared by a CVD method and transferred onto the end-face of a fiber connector to build a conventional soliton mode-locked fiber laser pulses operating at 1.57 um wavelength in anomalous dispersion regime as shown in Figure 9a–c [200].

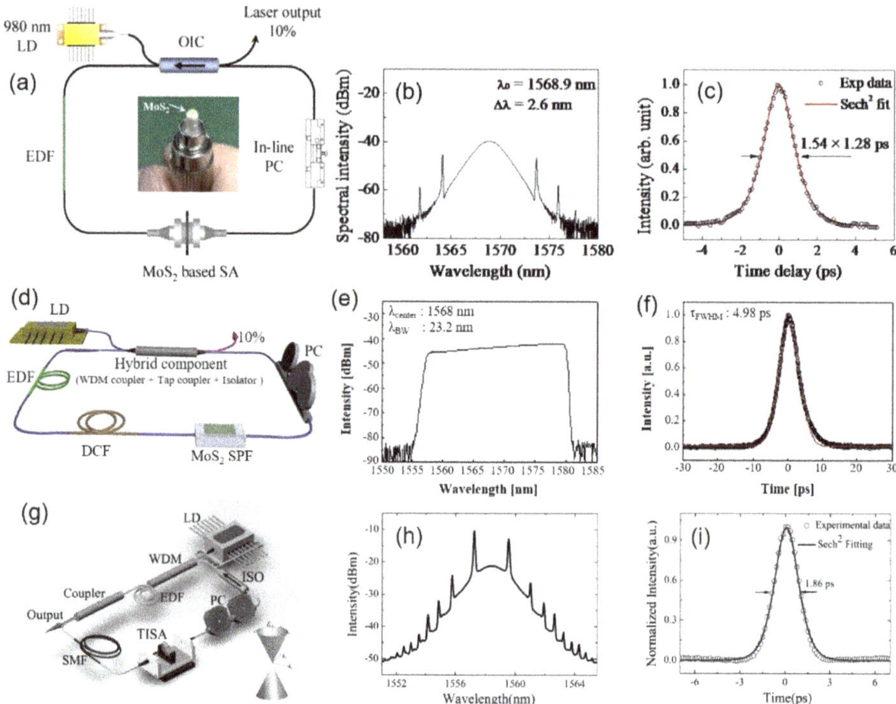

Figure 9. Mode-locked ultrafast fiber laser in the NIR region with various 2D SAs such as one of the TMDs named MoS_2 SA (**a–f**) and one of the TIs named Bi_2Te_3 (**g–i**). (**a–c**) CS with MoS_2 SA: (**a**) A schematic diagram of the fiber ring laser in anomalous dispersion with MoS_2 SA sandwiched in between two SMF (inset: Photograph of a fiber connector coated with multilayer MoS_2); (**b**) Optical spectrum of conventional soliton mode-locked pulse with a bandwidth of 2.6 nm; (**c**) Autocorrelation trace with a $Sech^2$ fit showing pulse duration of 1.28 ps. (**a–c**) Reproduced with permission. Copyright 2014, The Optical Society [200]. (**d–f**) DS with MoS_2 SA (**d**) A schematic diagram of the fiber ring laser in normal dispersion with MoS_2 SA deposited on SPF. (**e**) The optical spectrum of the mode-locked laser at net cavity dispersion of +0.095 ps^2 and (**f**) Measured pulsed duration fitted with Gaussian pulse. (**d–f**) Reproduced with permission. Copyright 2014, The Optical Society [162]. (**g–i**) CS with Bi_2Te_3 SA: (**g**) mode-locked fiber laser cavity comprising Bi_2Te_3 SA, (**h**) mode-locked optical spectrum with 3 dB bandwidth of 2.69 nm, and (**i**) autocorrelation trace of mode-locked soliton pulse with the FWHM width of 1.86 ps. (**g–i**) Reproduced with permission. Copyright 2012, AIP [201].

Resultant output soliton pulses showed central wavelength, spectral width, pulse duration, and repetition rate of 1568.9 nm, 2.6 nm, 1.28 ps, and 8.288 MHz, respectively. Khazaeizhad et al., in another work, successfully employed a CVD-grown multilayer MoS_2 SA in a passively mode-locked Er-doped fiber laser to achieve both soliton and dissipative soliton pulses [162]. Their dissipative soliton pulses characteristics operating at the near infrared region are shown in Figure 9d–f. The normal dispersion cavity was achieved by optimizing the net dispersion of the cavity by adding a segment of dispersion compensating fiber (DCF) in the cavity, as seen in Figure 9d. The stable dissipative soliton pulses with 4.98 ps pulse width at the repetition rate of 26.02 MHz showing a broad spectral width of 23.2 nm. In anomalous dispersion regime, they also obtained soliton pulses with pulse duration of 637 fs at the repetition rate of 33.48 MHz, and with a spectral width of 12.38 nm in an anomalous dispersion cavity. These findings have greatly promoted the development of few-layer MoS_2 in mode-locked lasers, leading to significant progress in this area.

As new LD graphene-like materials, TIs have been discovered with an energy band structure of symmetry Dirac cone due to their strong spin-orbit interaction, which implies that they can be developed into a new kind of SAs [202,203]. TIs have a nonzero bandgap and a large modulation depth (up to 95%), which are beneficial for improving the performance of mode-locked fiber lasers. Various mode-locked fiber lasers based on TIs, including Bi_2Se_3 [204,205], Bi_2Te_3 [107,108], and Sb_2Te_3 [53], have been developed, most of them are summarized in Table 1 with their mode-locked ultrafast EDFL. TIs were also confirmed as possessing excellent nonlinear optical properties and were used as SAs for demonstrating the ultrafast fiber laser in 2012 [201], as shown in Figure 9g–i. Zhao et al. reported in a first example among all TI that Bi_2Te_3 SA exhibited very-high-modulation-depth (up to 95%) saturable absorber and used as a passive mode locker for ultrafast pulse formation at the telecommunication band as shown in Figure 9g–i. In an erbium-doped fiber laser with the help of this SA, self-started mode-locked pulses centered at 1558.4 nm with a pulse width of 1.21 ps could be directly generated out of the laser cavity. These results indicate that, in addition to their established attractive electrical and thermal properties, TIs also have attractive application prospects for an ultrafast fiber laser.

All these LD SA devices also have been studied to build Q-switched or Q-switched mode-locked fiber laser by changing the cavity conditions such as polarization state, pump power or even by changing the fiber length of the cavity [67,68]. That means there is an additional regime of the fiber laser using these LD SA, including Q-switching and bound-state soliton mode-locking, to achieve high repetition rate fiber lasers in the gigahertz ranges useful for diverse applications. Therefore, the controllability of the fiber laser operating at different regimes using these LD SA would be of great interest due to its compatibility with various fiber-optic systems.

5. Externally Controlled Ultrafast Fiber Laser

To date, there are very few investigations reported on the active control of fiber laser using the LD SA, such as graphene and CNT [25–27,66–68]. These can be done both optically and/or electrically. It is noted that most of these active controls were done for evanescent field interaction cases, as it provides the facilitated device fabrication on the SPF surface along with the high functionality owing to the effective polarization sensitivity of evanescent field wave incident on SPF surfaces. It has been demonstrated that the fermi level of LD SA can be shifted by applying external gate bias to the LD-based capacitor structure, which plays a crucial role in tuning the modulation depth of the SA hence effectively control the fiber laser operating regime [25,67]. Apart from the evanescent field interaction cases, few attempts to simplify the design of actively mode-locked lasers have been made by employing a compact LD SA-based electro-optic modulator which controls the linear optical absorption as well as the modulation depth of the device upon applied external electrical signal [26,27]. These electro-optic modulators have been utilized in fiber laser for active control of pulse generation with tunable repetition rate along with mode-locking and harmonic mode-locking (HML) operation. As for the external optical bias, the nonlinear absorption in the SA can be substantially controlled via cross-absorption modulation (XAM) using evanescent field interaction; thus, the fiber laser operation was optically controlled [68]. A few of the recent works on externally controlled 2D SA for tunable fiber lasers and their characteristics are summarized in Table 2. The following section will briefly review those works covered by both electrically and optically controlled switchable fiber lasers.

Table 2. Performance summary of Externally controlled LD SA for tunable fiber laser.

LD SA (Type of Control)	Operating Voltage (V) or Control Beam Power (mW)	Fiber Laser Regime (QS or ML)	Repetition Rate (MHz)	Pulse Duration (ps)	3 dB Bandwidth (nm)	Center Wavelength (nm)	Ref.
Bi-layer graphene (Electrically controlled-ion-liquid gated)	-1.05 V -0.18 V	ML QS	30.9 25.4 kHz	0.423 3.5 µs	8 -	1609 1590	[25]
Bi-layer Graphene(Electrically controlled-PMMA)	±4 V -	ML QS-Not tested	2.44 -	0.390 -	8.9 -	1547.5 -	[66]
Mono-layer Graphene (Electro-optic modulator)	8 V (-4 to $+4$ V, 4.35 MHz) 8 V (-4 to $+4$ V, 8.70 MHz)	ML HML	4.35 8.70	1.44 1.57	1.8 1.82	1559.2 1559.3	[26]
Bi-layer Graphene (Electro-optic modulator)	3.5 V (35 to 65 KHz)	QS	35 to 65 KHz	1.93 to 5.54 µs	0.05	1524.6 to 1561.7	[27]
SWCNT (Electrically controlled-ion-liquid gated)	0~0.7 V 0.8~1.9 V	ML QS	50 23.6~28.8 kHz	0.6 -	7.6 -	1558 1559	[67]
Monolayer graphene (Optically controlled-ion-liquid gated), Biased by 980 nm CW pump beam	42 mW (980 nm) 34 mW (980 nm) 0 mW (980 nm) 30 mW Modulated signals Square pulse (1 ms)	ML QS-ML QS QS QS	5.09 20 kHz 5.09 MHz 8 kHz 6.2~11.8 kHz 1 Hz	980 - 20 µs - -	2.6 - - - -	1570 - - - -	[68]

5.1. Electrically Controlled Gate-Tunable Fiber Lasers

Lee et al., for the first time, demonstrated electrically controllable all-fiber graphene SA operating in Q-switching and mode-locking regime depending on the external gate-bias [25]. CVD grown, high-quality large-area graphene capacitor structure was fabricated on the SPF surface using a droplet of ion liquid on it as shown in Figure 10a. When the gate voltage is applied, the ions form an electric double layer (EDL) at the liquid/graphene interface with effective capacitance thickness around 1 nm over the entire graphene area. The exceptionally elevated capacitance of the electric double layer directs to an effective change in the Fermi level of graphene, which noticeably tunes the optical transmission of the incident light on the graphene SA at comparably lower operation voltage. As shown in Figure 10b, at zero gate bias voltage (V_G), the TE mode undergoes large absorption of 3.47 dB (>50%) (solid orange line at V_G = 0 V), while TM mode transmission rises by 0.09 dB (solid blue line at V_G = 0 V). This result is due to the reduced scattering loss caused by incorporating ion-liquid over-cladding on top of graphene and the polished surface.

More importantly, the applied gate voltage actively tuned the optical transmission of the device. TE mode exhibited a comparably larger change in transmission from 39.2 to 83.4% compared to the TM mode (from 87.1 to 90.8%) for V_G varying from 0.7 V to -1.8 V. This optical measurement implies that 44.2% of total incident light TE mode recovered after interacting with the single-layer graphene. These changes in optical transmission occur because of the Fermi level shift in graphene by an external gate voltage. Upon applying gate bias, the Fermi level near the Dirac point experiences substantial change due to incomplete density of state of electrons enclosed in a 2D graphene, revealing large electro-optic absorption. The corresponding change of electron carrier density in graphene was analyzed by determining the electrical transport properties of the device, as shown in Figure 10c. While the Fermi energy level reaches half the incident photon energy ($\hbar\omega/2$), the maximum transition in optical transmission appears to happen. The Fermi level change of 0.40 eV, which is the half-photon energy of the incident light source used in their experiment (1550 nm), is directly connected to the gate voltage deviation of 1.65 V, as described in Figure 10c. These integrated optical and electrical measurements and investigation could be a new technique to evaluate critical parameters like quantum capacitance and gate coupling efficiency without using complicated and expensive measurement systems. Nonlinear absorption characteristics of graphene such as saturation fluence and nonlinear modulation depth can be continuously tuned with the gate voltage, which can overcome the discrete nature of nonlinear absorption in stacked graphene sheets and be employed as functionally switchable devices. They also measured and observed the tunable nonlinear optical transmission as well as the varied modulation depth depending on the applied gate bias.

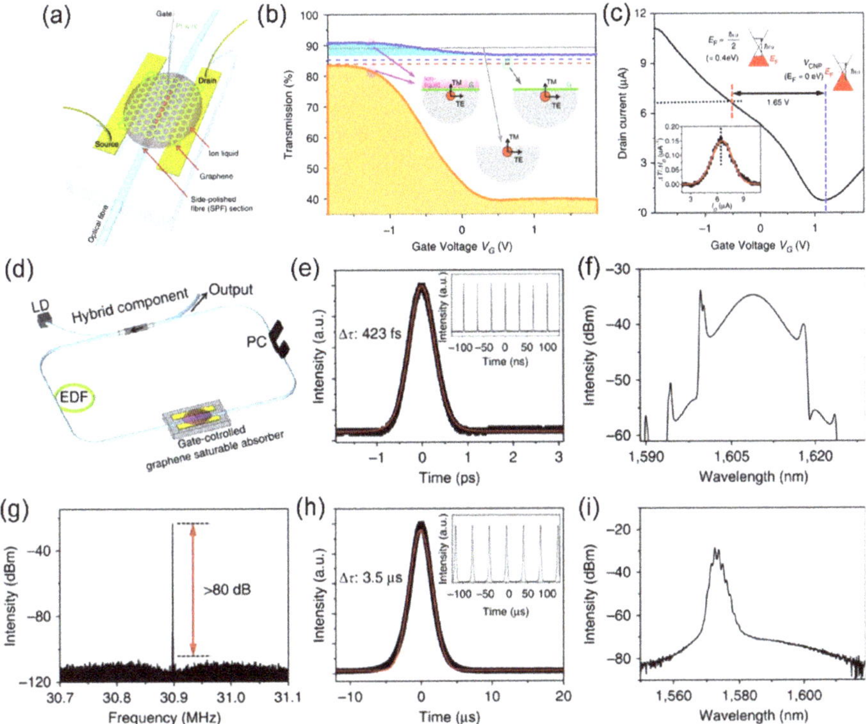

Figure 10. Electrically controlled fiber laser using an all-fiber graphene device and gate-variable properties of fiber laser operation. (**a**) Schematic diagram of gate-variable all-fiber graphene device. (**b**) Gate-controlled Optical transition properties of the device. (**c**) Gate-controlled Electrical transport properties of the device. (**d**) Fiber laser configuration, including fabricated all-fiber device with bilayer graphene. LD: laser diode; EDF: erbium-doped fiber; PC: polarization controller; hybrid component: an integrated wavelength-division multiplexer and isolator. (**e–g**) Characteristics of a passively mode-locked fiber laser at an applied V_G of -1.05 V; (**e**) Measured pulse duration of 423 fs at a repetition rate of 30.9 MHz (inset). (**f**) Laser output spectrum with a spectral bandwidth of 8 nm at 3 dB. (**g**) The measured radio frequency spectrum of the laser output (**h,i**) Q-switched characteristics of a fiber laser at an applied V_G of -0.18 V; (**h**) Measured output pulse duration of 3.5 ms at a repetition rate of 25.4 kHz (inset) and (**i**) its optical spectrum. Reproduced with permission. Copyright 2015, Springer Nature [25].

The fiber laser was built up using this device with a bilayer graphene SA case, which shows a self-starting of passive mode-locking operation at the applied V_G of -1.05 V where the measured pulse duration was 423 fs at a repetition rate of 30.9 MHz, as shown in Figure 10d–f. The spectral bandwidth of the laser output was measured to be 8.0 nm at the central wavelength of 1609 nm. The background noise level was >80 dB from the signal of the fundamental repetition rate of the laser output as shown in Figure 10g, which indicates stable mode-locking operation on increasing V_G, the linear optical transmission decreases, while both modulation depth and saturation fluence of the SA increase. This significantly modifies the Q-switching instability condition, changing the fiber laser operation to Q-switching. Figure 10h,i shows measured Q-switched pulse duration (3.5 μs) and optical spectrum, respectively, at the applied V_G of −0.18 V. The repetition rate of the laser was measured as 25.4 kHz (inset of Figure 10h). For an applied voltage larger than 0.14 V or less than −1.65 V, the fiber laser turns to continuous-wave operation.

Bogusławski et al. demonstrated an active mode-locked laser that achieved controlled ultrafast mode-locked laser pulses with tunable repetition rate by using a graphene-based electro-optic modulator (GEOM) as seen in Figure 11 [26]. The active mode-locking and active harmonic mode-locking of the erbium-doped fiber laser with output pulse duration of 1.44 ps and pulse energy of 844 pJ were achieved by the combination of the active mode-locking technique and the intracavity nonlinear pulse compression effect. The GEOM was integrated with an EDF laser working at 1.56 μm wavelength for active mode-locking. The experimental setup of the laser is presented in Figure 11a. The GEOM was inserted in a linear part of the cavity coupled with the ring-shaped part by a fiber circulator. A fiber collimator (L1) and an aspheric focusing lens (f = ≈4.6 mm, L2) were used to extract the beam outside the fiber and focus on the surface of the GEOM. The active mode-locking operation occurred immediately when the modulating signal frequency was set to precisely match the roundtrip frequency of the laser cavity (f_0 = 4.3505725 MHz) and the pump power exceeds the CW lasing threshold of 18 mW. The modulating electrical signal with the amplitude and frequency of 8 V (from −4 to 4 V) and 4.35 MHz respectively, was applied to GEOM as drive voltage to see the fundamental mode-locking operation with the repetition rate of 4.35 MHz as shown in the pulse train (Figure 11b). As soon as the modulation signal frequency was raised to 8.7012 MHz, which corresponds to second harmonic of the laser cavity, the second harmonic mode-locking (HML) operation occurred at 1559 nm with the pulse repetition rate of 8.7 MHz as shown in Figure 11c.

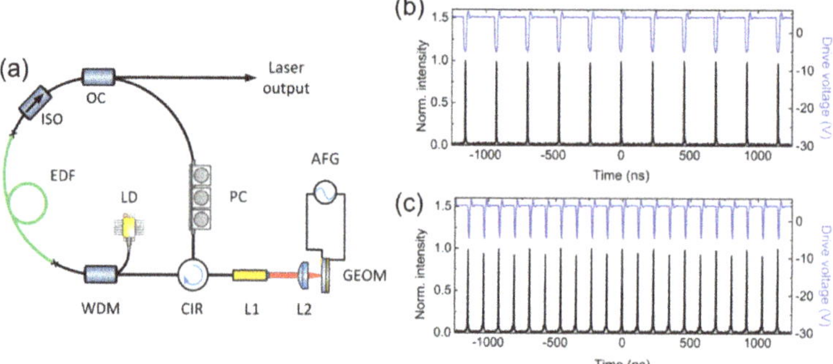

Figure 11. Graphene electro-optic modulator (GEOM) controlled mode-locked ultrafast fiber laser. (**a**) Laser cavity setup. EDF: erbium doped fiber; LD: laser diode; WDM: wavelength-division multiplexer; ISO: isolator; OC: output coupler; PC: polarization controller; CIR: circulator; L1: collimating lens; L2: focusing lens; AFG: arbitrary function generator. (**b**) Drive signal at the modulation frequency of 4.35 MHz and synchronized output optical pulse train for fundamental mode-locking operation at the repetition rate of 4.35 MHz. (**c**) Drive signal at modulation frequency of 4.35 MHz and synchronized output optical pulse train for second harmonic mode-locking operation at the repetition rate of 8.7 MHz. Reproduced with permission. Copyright 2018, Wiley-VCH [26].

Gladish et al., in another work, reported that electrochemical doping could tailor the nonlinear optical absorption of SWCNT films and demonstrated its application to control pulsed fiber laser generation in a similar manner [67]. The SWCNTs were manufactured by the aerosol CVD technique and followed by a dry transfer technique to a polarization-maintaining side-polished fiber (PM-SPF) to execution of the fiber laser system. They exhibited a comprehensive mode-locked ultrafast pulsed laser employing this device capable of operating in both mode-locked and Q-switched regimes manipulated by the external gate voltage. Self-starting of mode-locking happened as the pump power was boosted up to 40 mW. Stable mode-locked lasing at the fundamental pulse repetition rate of 50 MHz and the spectrum width at half-maximum is 7.6 nm with 0.6 ps pulse duration was obtained at zero gate voltage as shown in Figure 12a,b. If the applied gate voltage reached over the threshold voltage of 0.7 V, the ML pulse disappeared, and the cavity operation regime was switched to the Q-switched laser. Microsecond pulses generated in the QS regime exhibited repetition rates in kHz regime (Figure 12c). If the gate voltage was further increased up to 1.9 V, the Q-switched laser was persistent with an increasing repetition rate. The repetition rate of the Q-switched laser pulse was tuned from 23.6 to 28.8 kHz by increasing the applied gate voltage from 0.8 V to 1.9 V (Figure 12d). The pulse energy of the laser was recorded as high as 12.5 nJ.

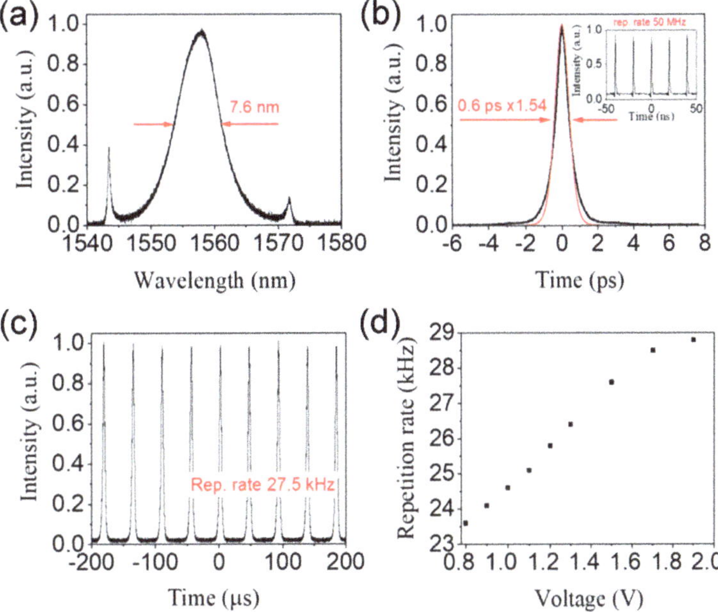

Figure 12. Electrically controlled fiber laser using SWCNT SA device and gate-variable properties of fiber laser operation. Soliton ML fiber laser characteristics at $V_G = 0$ V: (**a**) optical soliton pulse spectrum shows the 3 dB bandwidth of 7.6 nm, (**b**) autocorrelation trace shows the pulse duration of 600 fs with oscillation trace (inset showing repetition rate of 50 MHz) of the mode-locked pulse. QS fiber laser characteristics at $V_G > 0.7$ V: (**c**) oscillation trace of QS pulse with repetition rate of 27.5 kHz, (**d**) variation of QS pulse frequency as a function of applied gate voltage showing that, repetition rate is controlled in the range from 23.6 kHz to 28.8 kHz with applied gate voltage V_G varying from 0.8 V to 1.9 V. Reproduced with permission. Copyright 2019, American Chemical Society [67].

5.2. Optically Controlled Cross-Absorption Modulated Tunable Fiber Laser

Apart from the electrical control, optically excited carriers also alternatively tune the linear optical absorption in graphene through Pauli blocking theory, which facilitated graphene-based devices for all-optical modulation in broadband scale [206]. This can adjust the optical properties of graphene SA on a nonlinear scale, such as modulation depth and non-saturable loss. This method was utilized in the fiber laser cavity by Sheng et al. to optically manipulate and tune the pulse width of a passively mode-locked ultrafast fiber laser based on graphene SA [207]. Cross absorption modulation (XAM) employing evanescent field interaction significantly modified the nonlinear SA in graphene. Gene et al., in the recent past, successfully accomplished different pulsed fiber laser regimes by optically controlling and manipulating the SA [68]. They displayed optically tuned in-line graphene SA with higher tunability in modulation depth by means of boosting the interaction of graphene SA with the incident evanescent field. The all-fiber graphene SA was manufactured by transferring a uniform and large-area single-layer graphene onto the side-polished fiber (SPF), as seen in Figure 13a. Nonlinear absorption properties of graphene SA were optically controlled by means of XAM in a graphene sheet. A shorter wavelength (980 nm) light was utilized to manipulate the absorption of the signal beam at a higher wavelength (1550 nm). As the graphene SA layer absorbed the control beam, the signal beam experienced less absorption by XAM [207]. In the case of the parallel polarization direction of TE mode along with the graphene layer at a low power level, the incident signal beam encounters only a slight transmission of 2.5% in the monolayer graphene as shown in Figure 13b. At a high-power level, 50.2% of the transmission recorded in the absence of the applied control beam reveals the modulation depth of 47.7% over the 800 mW of signal power. The modulation depth reduced to 28.5% as soon as the incident control beam power reached 84 mW. The graphene SA device was incorporated into an Er-doped fiber ring-laser system, as illustrated in Figure 13c. The laser cavity comprises two wavelength division multiplexing (WDM) couplers and two 980-nm laser diodes (LDs), pumping the EDF and the other to control the graphene SA. A pulsed fiber laser is capable of operating in several modes such as Q-switching, Q-switched mode-locking (QML), and CW mode-locking mode, which can be governed by manipulating the system parameters such as modulation depth, gain relaxation time, gain saturation power, small-signal gain, and saturation power of the SA. Figure 13d shows the switching performance of the fiber laser regime as a function of SA modulation depth. Figure 13e–g illustrates experimental findings of the switching actions altered between different laser operating regimes achieved through active control graphene SA by manipulating the external control beam in a pulsed fiber laser setup. In the absence of a control beam (P_c = 0 mW), graphene SA in the ring cavity reveals the Q-switched laser (Figure 13e), with the pulse duration and repetition rate of 20μs and 8.0 kHz, respectively. The laser state was switched to QML if the control beam power was elevated to 34 mW. A fine mode-locked pulse train at a repetition rate of 5.09 MHz (inset of Figure 13f) is clearly observed within the 20 kHz Q-switched pulses, as shown in Figure 13f. CW mode-locking status was achieved as the control beam power was increased to 42 mW (Figure 13g). In addition to these, they also examined the laser characteristics when a control beam with sinusoidal modulation frequency was applied to the graphene SA at a particular power of the control beam. It was seen that the generated Q-switched laser emitted from the cavity exhibited the repetition rate same as the modulation speed of the applied control beam. A Q-switched laser with a continuously tunable repetition rate in the range from 6.2 kHz to 11.8 kHz was obtained by manipulating the modulation frequency of the control beam. The controllable and switchable fiber pulse laser point towards a suitable solution for developing versatile pulse laser seed sources. This kind of source can potentially find a number of applications both in Q-switching and mode-locking regimes (for example, remote sensing, distance measurement, time-resolved spectroscopy, optical frequency metrology, or photo-acoustic imaging). This concept can be further expanded to an optical fiber taper that potentially provides more tight confinement of light, resulting in reduced device size with the higher operating speed of the device. The proposed scheme can also be applied to the plastic optical waveguides combined with an ionic gel, paving a novel way for actively controlled, flexible photonic devices. Based on this and combined with the fundamental studies, high repetition rate ultrafast fiber lasers or tunable repetition rate Q-switched fiber lasers also can be considered for future works. These investigations of graphene and CNT for externally controlled fiber lasers are also exciting to explore other LD SAs, especially 2D SAs, to find efficient optical devices with high performance and precisely tunable functionalities.

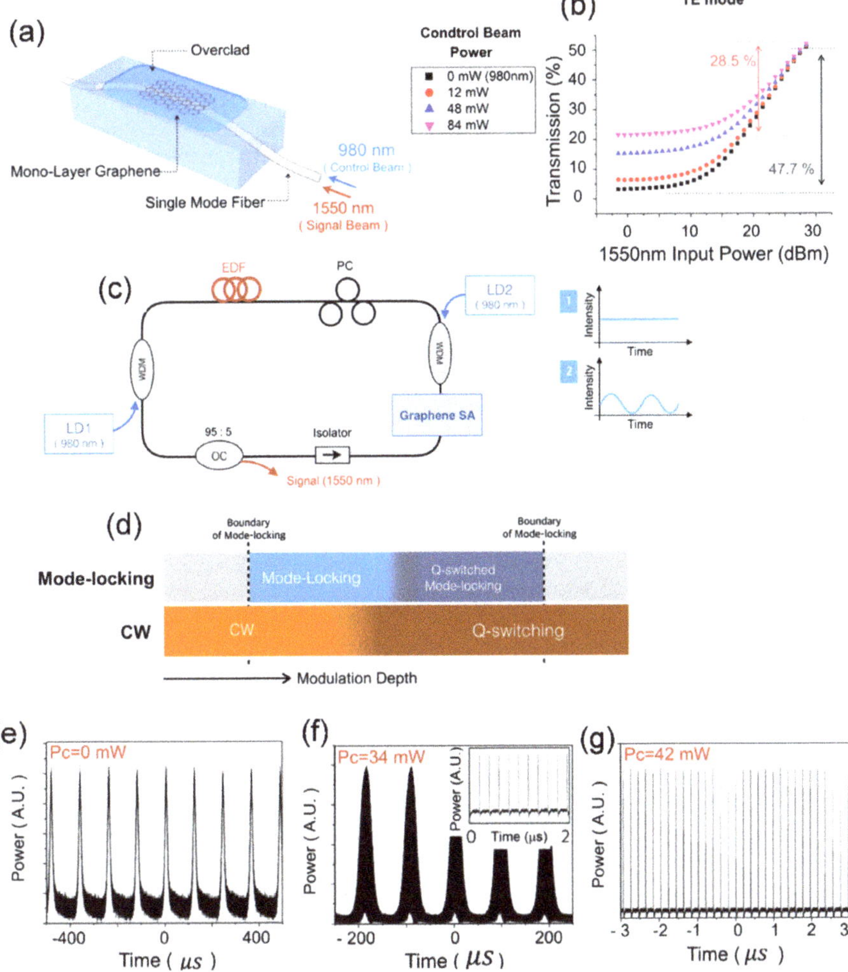

Figure 13. Optically controlled in-line graphene SA-based pulsed fiber laser (**a**) Schematic representation of optically tunable graphene SA. (**b**) Nonlinear transmission test result of the CW signal beam (1550 nm) TE mode variable with CW control beam powers at 980 nm. (**c**) Illustration of Er-doped fiber ring laser incorporated with optically controllable in-line monolayer graphene SA device. LD: laser diode; EDF: erbium-doped fiber; PC: polarization controller; WDM: wavelength-division multiplexer and OC: optical coupler. (**d**) Schematic explanation of fiber laser operating regime as a function of modulation depth in graphene SA, (**e**) Q-switching operated pulse train with no control beam (P_c = 0 mW) applied, (**f**) Q-switched mode-locked operated pulse train with a control beam power of 34 mW (inset: an extended view of the pulse train in time scale) and (**g**) pulse train of CW mode-locked operation with a control beam power of 42 mW. Reproduced with permission. Copyright 2016, OSA [68].

6. Prospects for Future Research Directions

Low-dimensional materials, particularly 2D materials, are a rising and hot topic expanding in quality, variety, and quantity. Over the last decade, the development of pulsed lasers based on LD materials has progressed rapidly, yielding many significant results and being employed in various applications. This is due to the reliability of synthesis of LD materials, coupling technology to implement SA devices, and the steady advancement of pulsed fiber laser technology over the decades. Nevertheless, there are still numerous challenges to realize high performance and finely tunable pulsed fiber laser and technical as well as scientific issues to be solved. Controllability of fiber laser in between Q-switching and mode-locking has been achieved by electrically and/or optically tuning the Fermi level, the modulation depth, and cross absorption modulation of LD materials. However, in the case of ion-liquid over-cladding, the switching speed and stability are limited by the ionic mobility in the ion liquid. This prospects towards the development of stable over-cladding in the capacitor structure based on LD SA on SPF that will benefit the stability and specifically control the mode-locking regime. Numerous stable over-cladding indexes and solid polymer indexes have been developed to study and analyze the LD SA-based mode-locked fiber laser and photonics characteristics. However, the externally controlled or gate-tunable LD SA-based fiber laser has not been studied yet, employing the stable over-cladding index required for enhanced coupling as well as for the EDL formation, which is crucial for the Fermi level tuning in the LD SAs. Moreover, only 2D graphene and 1D CNT among a large variety of LD SAs has been investigated for a tunable mode-locked fiber laser. There are many other LD SAs to be analyzed in this specific field for a stable, high-performance controllable fiber laser. Additionally, the basic characteristics of the ultrafast fiber laser, such as repetition rate and pulse duration, may be effectively controlled utilizing the tunable behavior of LD SA upon external (electrical and optical) bias. All of these LD-based tunable devices could also be investigated for various in-line all-fiber devices, such as ultrafast all-optical tunable switchers, optical limiters, all-optical modulators, polarizers etc., based on electrically/optically controlled nonlinear optical properties (i.e., higher-order susceptibility/nonlinearity, multiphoton absorption, etc.). The twist-angle in Bi-layer LD materials also can be engineered along with the externally (electrically and/or optically) controlled scheme to find the enhanced higher-order optical nonlinearity and SA properties for high performance, ultrafast and precisely controlled fiber lasers, and other nonlinear optical devices [54]. LD SA-based fs mode-locked ultrafast fiber lasers have grown emerging interest in space-borne applications, as they can sustain more than the life span of satellites [136,208,209]. Other radiation environments such as particle accelerators and radiation-based medical instruments also could be benefit from utilizing the controllable functionality of ultrafast fiber lasers based on actively controlled characteristics of LD SAs. The tunability of such fiber lasers could provide significant functionality in that application for the development of future space technology and many other advanced applications.

7. Conclusions

In conclusion, we present a brief review of numerous LD SA-based ultrafast fiber lasers and their performance reported so far in various technical schemes. Based on the aforementioned review, we estimate that LD materials with broadband optical response, high stability, good reliability, excellent thermal performance, low defects, and precisely controllable properties, which are compatible for high-energy ultrafast fiber laser along with the functional operation, will be constructed and taken out of the laboratory for real-world applications. Although there have been several reports on the electrically controlled nonlinear transmission and saturable absorption in graphene and carbon nanotube-based mode-locked ultrafast fiber laser and Q-switched laser, there is still more room to explore externally controlled saturable absorbers based on other LD SAs. The electrical and optical gating of graphene and SWCNT exposes the opportunities for the development of externally (electrically or optically) tunable other various LD-materials based nonlinear optical devices and points towards advanced device performances with tunable nonlinearity and controllable functionalities. Principally in both practical and theoretical frameworks, the externally (electrically and/or optically) tunable nonlinear optical operation of 2D graphene and other LD nonlinear optical materials also propose a variety of technical and scientific benefits, such as devices with a compact minimum footprint, exceedingly fast speed (more than a few tens of GHz), chip-scale integration and compatibility with complementary metal-oxide-semiconductor (CMOS) technology, all of which are required ideal standards for future on-chip photonic and optoelectronic applications. These will also benefit the basic study and investigation of LD materials for further understanding the incomparable advantages of nonlinear optical as well as ultrafast photonic systems over their electronic counterparts.

Author Contributions: Conceptualization, P.C.D. and D.-I.Y., Methodology, P.C.D. and D.-I.Y., Data collection P.C.D., Writing P.C.D. and D.-I.Y. All authors have read and agreed to the published version of the manuscript.

Funding: This work was supported by Brain Pool Program through the National Research Foundation of Korea (NRF) funded by the Ministry of Science and ICT (NRF-2019H1D3A1A02071061).

Institutional Review Board Statement: Not applicable.

Informed Consent Statement: Not applicable.

Data Availability Statement: Data is contained within the article which are from various articles cited as references.

Conflicts of Interest: The authors declare no conflict of interest.

References

1. Chou, S.Y.; Keimel, C.; Gu, J. Ultrafast and direct imprint of nanostructures in silicon. *Nature* **2002**, *417*, 835–837. [CrossRef]
2. Ma, J.; Xie, G.Q.; Gao, W.L.; Yuan, P.; Qian, L.J.; Yu, H.H.; Zhang, H.J.; Wang, J.Y. Diode-pumped mode-locked femtosecond Tm:CLNGG disordered crystal laser. *Opt. Lett.* **2012**, *37*, 1376–1378. [CrossRef]
3. Kondo, Y.; Nouchi, K.; Mitsuyu, T.; Watanabe, M.; Kazansky, P.G.; Hirao, K. Fabrication of long-period fiber gratings by focused irradiation of infrared femtosecond laser pulses. *Opt. Lett.* **1999**, *24*, 646–648. [CrossRef]
4. Marcinkevi Ius, A.; Juodkazis, S.; Watanabe, M.; Miwa, M.; Matsuo, S.; Misawa, H.; Nishii, J. Femtosecond laser-assisted three-dimensional microfabrication in silica. *Opt. Lett.* **2001**, *26*, 277–279. [CrossRef]
5. Schaffer, C.B.; Brodeur, A.; García, J.F.; Mazur, E. Micromachining bulk glass by use of femtosecond laser pulses with nanojoule energy. *Opt. Lett.* **2001**, *26*, 93–95. [CrossRef]
6. Keller, U. Recent developments in compact ultrafast lasers. *Nature* **2003**, *424*, 831–838. [CrossRef]
7. Fermann, M.E.; Hartl, I. Ultrafast fibre lasers. *Nat. Photonics* **2013**, *7*, 868–874. [CrossRef]
8. Guo, B.; Xiao, Q.-l.; Wang, S.-h.; Zhang, H. 2D Layered Materials: Synthesis, Nonlinear Optical Properties, and Device Applications. *Laser Photonics Rev.* **2019**, *13*, 1800327. [CrossRef]
9. He, J.; Tao, L.; Zhang, H.; Zhou, B.; Li, J. Emerging 2D materials beyond graphene for ultrashort pulse generation in fiber lasers. *Nanoscale* **2019**, *11*, 2577–2593. [CrossRef] [PubMed]
10. Okhotnikov, O.; Jouhti, T.; Konttinen, J.; Karirinne, S.; Pessa, M. 1.5-μm monolithic GaInNAs semiconductor saturable-absorber mode locking of an erbium fiber laser. *Opt. Lett.* **2003**, *28*, 364–366. [CrossRef] [PubMed]
11. Aguergaray, C.; Broderick, N.G.R.; Erkintalo, M.; Chen, J.S.Y.; Kruglov, V. Mode-locked femtosecond all-normal all-PM Yb-doped fiber laser using a nonlinear amplifying loop mirror. *Opt. Express* **2012**, *20*, 10545–10551. [CrossRef]
12. Liu, X.; Zhan, L.; Luo, S.; Gu, Z.; Liu, J.; Wang, Y.; Shen, Q. Multiwavelength erbium-doped fiber laser based on a nonlinear amplifying loop mirror assisted by un-pumped EDF. *Opt. Express* **2012**, *20*, 7088–7094. [CrossRef] [PubMed]
13. Matsas, V.; Newson, T.; Richardson, D.; Payne, D.N. Self-starting, passively mode-locked fibre ring soliton laser exploiting non-linear polarisation rotation. *Electron. Lett.* **1992**, *28*, 1391–1393. [CrossRef]
14. Wei, C.; Shi, H.; Luo, H.; Zhang, H.; Lyu, Y.; Liu, Y. 34 nm-wavelength-tunable picosecond Ho^{3+}/Pr^{3+}-codoped ZBLAN fiber laser. *Opt. Express* **2017**, *25*, 19170–19178. [CrossRef] [PubMed]
15. Gluth, A.; Wang, Y.; Petrov, V.; Paajaste, J.; Suomalainen, S.; Härkönen, A.; Guina, M.; Steinmeyer, G.; Mateos, X.; Veronesi, S. GaSb-based SESAM mode-locked Tm: YAG ceramic laser at 2 μm. *Opt. Express* **2015**, *23*, 1361–1369. [CrossRef] [PubMed]
16. Keller, U.; Weingarten, K.J.; Kartner, F.X.; Kopf, D.; Braun, B.; Jung, I.D.; Fluck, R.; Honninger, C.; Matuschek, N.; Der Au, J.A. Semiconductor saturable absorber mirrors (SESAM's) for femtosecond to nanosecond pulse generation in solid-state lasers. *IEEE J. Sel. Top. Quantum Electron.* **1996**, *2*, 435–453. [CrossRef]
17. Keller, U.; Miller, D.; Boyd, G.; Chiu, T.; Ferguson, J.; Asom, M. Solid-state low-loss intracavity saturable absorber for Nd: YLF lasers: An antiresonant semiconductor Fabry–Perot saturable absorber. *Opt. Lett.* **1992**, *17*, 505–507. [CrossRef]
18. Okhotnikov, O.; Grudinin, A.; Pessa, M. Ultra-fast fibre laser systems based on SESAM technology: New horizons and applications. *New J. Phys.* **2004**, *6*, 177. [CrossRef]
19. Tang, P.; Qin, Z.; Liu, J.; Zhao, C.; Xie, G.; Wen, S.; Qian, L. Watt-level passively mode-locked Er3+-doped ZBLAN fiber laser at 2.8 μm. *Opt. Lett.* **2015**, *40*, 4855–4858. [CrossRef]
20. Sun, Z.; Hasan, T.; Torrisi, F.; Popa, D.; Privitera, G.; Wang, F.; Bonaccorso, F.; Basko, D.M.; Ferrari, A.C. Graphene mode-locked ultrafast laser. *ACS Nano* **2010**, *4*, 803–810. [CrossRef]

21. Bao, Q.; Zhang, H.; Wang, Y.; Ni, Z.; Yan, Y.; Shen, Z.X.; Loh, K.P.; Tang, D.Y. Atomic-layer graphene as a saturable absorber for ultrafast pulsed lasers. *Adv. Funct. Mater.* **2009**, *19*, 3077–3083. [CrossRef]
22. Yamashita, S.; Inoue, Y.; Maruyama, S.; Murakami, Y.; Yaguchi, H.; Jablonski, M.; Set, S. Saturable absorbers incorporating carbon nanotubes directly synthesized onto substrates and fibers and their application to mode-locked fiber lasers. *Opt. Lett.* **2004**, *29*, 1581–1583. [CrossRef] [PubMed]
23. Hasan, T.; Sun, Z.; Wang, F.; Bonaccorso, F.; Tan, P.H.; Rozhin, A.G.; Ferrari, A.C. Nanotube–polymer composites for ultrafast photonics. *Adv. Mater.* **2009**, *21*, 3874–3899. [CrossRef]
24. Du, J.; Zhang, M.; Guo, Z.; Chen, J.; Zhu, X.; Hu, G.; Peng, P.; Zheng, Z.; Zhang, H. Phosphorene quantum dot saturable absorbers for ultrafast fiber lasers. *Sci. Rep.* **2017**, *7*, 42357. [CrossRef] [PubMed]
25. Lee, E.J.; Choi, S.Y.; Jeong, H.; Park, N.H.; Yim, W.; Kim, M.H.; Park, J.K.; Son, S.; Bae, S.; Kim, S.J.; et al. Active control of all-fibre graphene devices with electrical gating. *Nat. Commun.* **2015**, *6*, 6851. [CrossRef] [PubMed]
26. Bogusławski, J.; Wang, Y.; Xue, H.; Yang, X.; Mao, D.; Gan, X.; Ren, Z.; Zhao, J.; Dai, Q.; Soboń, G.; et al. Graphene Actively Mode-Locked Lasers. *Adv. Funct. Mater.* **2018**, *28*, 1801539. [CrossRef]
27. Li, D.; Xue, H.; Qi, M.; Wang, Y.; Aksimsek, S.; Chekurov, N.; Kim, W.; Li, C.; Riikonen, J.; Ye, F.; et al. Graphene actively Q-switched lasers. *2D Mater.* **2017**, *4*, 025095. [CrossRef]
28. Lee, J.; Koo, J.; Chang, Y.M.; Debnath, P.; Song, Y.-W.; Lee, J.H. Experimental investigation on a Q-switched, mode-locked fiber laser based on the combination of active mode locking and passive Q switching. *J. Opt. Soc. Am. B* **2012**, *29*, 1479–1485. [CrossRef]
29. Xia, H.; Li, H.; Lan, C.; Li, C.; Zhang, X.; Zhang, S.; Liu, Y. Erbium-doped fiber laser mode-locked with a few-layer MoS$_2$ saturable absorber. In Proceedings of the Asia Communications and Photonics Conference, Optical Society of America, Shanghai, China, 11–14 November 2014; p. ATh3A. 89.
30. Liu, H.; Zheng, X.W.; Liu, M.; Zhao, N.; Luo, A.P.; Luo, Z.C.; Xu, W.C.; Zhang, H.; Zhao, C.J.; Wen, S.C. Femtosecond pulse generation from a topological insulator mode-locked fiber laser. *Opt. Express* **2014**, *22*, 6868–6873. [CrossRef] [PubMed]
31. Bonaccorso, F.; Sun, Z.; Hasan, T.; Ferrari, A. Graphene photonics and optoelectronics. *Nat. Photonics* **2010**, *4*, 611. [CrossRef]
32. Debnath, P.C.; Park, J.; Scott, A.M.; Lee, J.; Lee, J.H.; Song, Y.W. In situ synthesis of graphene with telecommunication lasers for nonlinear optical devices. *Adv. Opt. Mater.* **2015**, *3*, 1264–1272. [CrossRef]
33. Jhon, Y.I.; Koo, J.; Anasori, B.; Seo, M.; Lee, J.H.; Gogotsi, Y.; Jhon, Y.M. Metallic MXene saturable absorber for femtosecond mode-locked lasers. *Adv. Mater.* **2017**, *29*, 1702496. [CrossRef] [PubMed]
34. Yi, Y.; Sun, Z.; Li, J.; Chu, P.K.; Yu, X.F. Optical and optoelectronic properties of black phosphorus and recent photonic and optoelectronic applications. *Small Methods* **2019**, *3*, 1900165. [CrossRef]
35. Debnath, P.C.; Park, K.; Song, Y.W. Recent Advances in Black-Phosphorus-Based Photonics and Optoelectronics Devices. *Small Methods* **2018**, *2*, 1700315. [CrossRef]
36. Lee, D.; Park, K.; Debnath, P.C.; Kim, I.; Song, Y.W. Thermal damage suppression of a black phosphorus saturable absorber for high-power operation of pulsed fiber lasers. *Nanotechnology* **2016**, *27*, 365203. [CrossRef] [PubMed]
37. Aktürk, E.; Aktürk, O.Ü.; Ciraci, S. Single and bilayer bismuthene: Stability at high temperature and mechanical and electronic properties. *Phys. Rev. B* **2016**, *94*, 014115. [CrossRef]
38. Dresselhaus, M.S.; Dresselhaus, G.; Eklund, P.C. *Science of Fullerenes and Carbon Nanotubes: Their Properties and Applications*; Elsevier: Amsterdam, The Netherlands, 1996.
39. Mak, K.F.; Lee, C.; Hone, J.; Shan, J.; Heinz, T.F. Atomically Thin MoS$_2$: A New Direct-Gap Semiconductor. *Phys. Rev. Lett.* **2010**, *105*, 136805. [CrossRef]
40. Nair, R.R.; Blake, P.; Grigorenko, A.N.; Novoselov, K.S.; Booth, T.J.; Stauber, T.; Peres, N.M.; Geim, A.K. Fine structure constant defines visual transparency of graphene. *Science* **2008**, *320*, 1308. [CrossRef]
41. Geim, A.K. Graphene: Status and Prospects. *Science* **2009**, *324*, 1530–1534. [CrossRef]
42. Geim, A.K.; Novoselov, K.S. The rise of graphene. *Nat. Mater.* **2007**, *6*, 183–191. [CrossRef]
43. Novoselov, K.S.; Fal, V.; Colombo, L.; Gellert, P.; Schwab, M.; Kim, K. A roadmap for graphene. *Nature* **2012**, *490*, 192–200. [CrossRef] [PubMed]
44. Novoselov, K.S.; Geim, A.K.; Morozov, S.V.; Jiang, D.; Katsnelson, M.I.; Grigorieva, I.; Dubonos, S.; Firsov, A. Two-dimensional gas of massless Dirac fermions in graphene. *Nature* **2005**, *438*, 197–200. [CrossRef]
45. Novoselov, K.S.; Geim, A.K.; Morozov, S.V.; Jiang, D.; Zhang, Y.; Dubonos, S.V.; Grigorieva, I.V.; Firsov, A.A. Electric field effect in atomically thin carbon films. *Science* **2004**, *306*, 666–669. [CrossRef] [PubMed]
46. Lu, L.; Wang, W.; Wu, L.; Jiang, X.; Xiang, Y.; Li, J.; Fan, D.; Zhang, H. All-Optical Switching of Two Continuous Waves in Few Layer Bismuthene Based on Spatial Cross-Phase Modulation. *ACS Photonics* **2017**, *4*, 2852–2861. [CrossRef]
47. Singh, D.; Gupta, S.K.; Sonvane, Y.; Lukačević, I. Antimonene: A monolayer material for ultraviolet optical nanodevices. *J. Mater. Chem. C* **2016**, *4*, 6386–6390. [CrossRef]
48. Lu, S.B.; Miao, L.L.; Guo, Z.N.; Qi, X.; Zhao, C.J.; Zhang, H.; Wen, S.C.; Tang, D.Y.; Fan, D.Y. Broadband nonlinear optical response in multi-layer black phosphorus: An emerging infrared and mid-infrared optical material. *Opt. Express* **2015**, *23*, 11183–11194. [CrossRef]
49. Wang, Q.H.; Kalantar-Zadeh, K.; Kis, A.; Coleman, J.N.; Strano, M.S. Electronics and optoelectronics of two-dimensional transition metal dichalcogenides. *Nat. Nanotechnol.* **2012**, *7*, 699–712. [CrossRef]

50. Hanlon, D.; Backes, C.; Doherty, E.; Cucinotta, C.S.; Berner, N.C.; Boland, C.; Lee, K.; Harvey, A.; Lynch, P.; Gholamvand, Z.; et al. Liquid exfoliation of solvent-stabilized few-layer black phosphorus for applications beyond electronics. *Nat. Commun.* **2015**, *6*, 8563. [CrossRef]

51. Koppens, F.H.L.; Mueller, T.; Avouris, P.; Ferrari, A.C.; Vitiello, M.S.; Polini, M. Photodetectors based on graphene, other two-dimensional materials and hybrid systems. *Nat. Nanotechnol.* **2014**, *9*, 780–793. [CrossRef]

52. Wu, K.; Wang, Y.; Qiu, C.; Chen, J. Thermo-optic all-optical devices based on two-dimensional materials. *Photonics Res.* **2018**, *6*, C22–C28. [CrossRef]

53. Wu, K.; Chen, B.; Zhang, X.; Zhang, S.; Guo, C.; Li, C.; Xiao, P.; Wang, J.; Zhou, L.; Zou, W. High-performance mode-locked and Q-switched fiber lasers based on novel 2D materials of topological insulators, transition metal dichalcogenides and black phosphorus: Review and perspective. *Opt. Commun.* **2018**, *406*, 214–229. [CrossRef]

54. Ha, S.; Park, N.H.; Kim, H.; Shin, J.; Choi, J.; Park, S.; Moon, J.-Y.; Chae, K.; Jung, J.; Lee, J.-H. Enhanced third-harmonic generation by manipulating the twist angle of bilayer graphene. *Light Sci. Appl.* **2021**, *10*, 1–10. [CrossRef] [PubMed]

55. Hsu, W.-T.; Zhao, Z.-A.; Li, L.-J.; Chen, C.-H.; Chiu, M.-H.; Chang, P.-S.; Chou, Y.-C.; Chang, W.-H. Second harmonic generation from artificially stacked transition metal dichalcogenide twisted bilayers. *ACS Nano* **2014**, *8*, 2951–2958. [CrossRef] [PubMed]

56. Li, J.; Zhang, Z.; Du, L.; Miao, L.; Yi, J.; Huang, B.; Zou, Y.; Zhao, C.; Wen, S. Highly stable femtosecond pulse generation from a MXene $Ti_3C_2T_x$ (T = F, O, or OH) mode-locked fiber laser. *Photonics Res.* **2019**, *7*, 260–264. [CrossRef]

57. Hao, Q.; Wang, C.; Liu, W.; Liu, X.; Liu, J.; Zhang, H. Low-dimensional saturable absorbers for ultrafast photonics in solid-state bulk lasers: Status and prospects. *Nanophotonics* **2020**, *9*, 2603–2639. [CrossRef]

58. Wang, J.; Wang, X.; Lei, J.; Ma, M.; Wang, C.; Ge, Y.; Wei, Z. Recent advances in mode-locked fiber lasers based on two-dimensional materials. *Nanophotonics* **2020**, *9*, 2315–2340. [CrossRef]

59. Im, J.H.; Choi, S.Y.; Rotermund, F.; Yeom, D.I. All-fiber Er-doped dissipative soliton laser based on evanescent field interaction with carbon nanotube saturable absorber. *Opt. Express* **2010**, *18*, 22141–22146. [CrossRef]

60. Jeong, H.; Choi, S.Y.; Rotermund, F.; Cha, Y.-H.; Jeong, D.-Y.; Yeom, D.-I. All-fiber mode-locked laser oscillator with pulse energy of 34 nJ using a single-walled carbon nanotube saturable absorber. *Opt. Express* **2014**, *22*, 22667–22672. [CrossRef]

61. Jeong, H.; Choi, S.Y.; Rotermund, F.; Lee, K.; Yeom, D.-I. All-polarization maintaining passively mode-locked fiber laser using evanescent field interaction with single-walled carbon nanotube saturable absorber. *J. Lightwave Technol.* **2016**, *34*, 3510–3514. [CrossRef]

62. Jeong, H.; Yeom, D.-I. Passively Q-switched Erbium Doped All-fiber Laser with High Pulse Energy Based on Evanescent Field Interaction with Single-walled Carbon Nanotube Saturable Absorber. *Curr. Opt. Photonics* **2017**, *1*, 203–206.

63. Jeong, H.; Choi, S.Y.; Rotermund, F.; Yeom, D.I. Pulse width shaping of passively mode-locked soliton fiber laser via polarization control in carbon nanotube saturable absorber. *Opt. Express* **2013**, *21*, 27011–27016. [CrossRef]

64. Jeong, H.; Choi, S.Y.; Jeong, E.I.; Cha, S.J.; Rotermund, F.; Yeom, D.-I. Ultrafast mode-locked fiber laser using a waveguide-type saturable absorber based on single-walled carbon nanotubes. *Appl. Phys. Express* **2013**, *6*, 052705. [CrossRef]

65. Jiang, T.; Yin, K.; Wang, C.; You, J.; Ouyang, H.; Miao, R.; Zhang, C.; Wei, K.; Li, H.; Chen, H.; et al. Ultrafast fiber lasers mode-locked by two-dimensional materials: Review and prospect. *Photonics Res.* **2020**, *8*, 78–90. [CrossRef]

66. Kovalchuk, O.; Uddin, S.; Lee, S.; Song, Y.-W. Graphene Capacitor-Based Electrical Switching of Mode-Locking in All-Fiberized Femtosecond Lasers. *ACS Appl. Mater. Interfaces* **2020**, *12*, 54005–54011. [CrossRef] [PubMed]

67. Gladush, Y.; Mkrtchyan, A.A.; Kopylova, D.S.; Ivanenko, A.; Nyushkov, B.; Kobtsev, S.; Kokhanovskiy, A.; Khegai, A.; Melkumov, M.; Burdanova, M. Ionic liquid gated carbon nanotube saturable absorber for switchable pulse generation. *Nano Lett.* **2019**, *19*, 5836–5843. [CrossRef] [PubMed]

68. Gene, J.; Park, N.H.; Jeong, H.; Choi, S.Y.; Rotermund, F.; Yeom, D.I.; Kim, B.Y. Optically controlled in-line graphene saturable absorber for the manipulation of pulsed fiber laser operation. *Opt. Express* **2016**, *24*, 21301–21307. [CrossRef] [PubMed]

69. Kong, L.; Qin, Z.; Xie, G.; Guo, Z.; Zhang, H.; Yuan, P.; Qian, L. Black phosphorus as broadband saturable absorber for pulsed lasers from 1 μm to 2.7 μm wavelength. *Laser Phys. Lett.* **2016**, *13*, 045801. [CrossRef]

70. Jiang, Y.; Miao, L.; Jiang, G.; Chen, Y.; Qi, X.; Jiang, X.-f.; Zhang, H.; Wen, S. Broadband and enhanced nonlinear optical response of MoS_2/graphene nanocomposites for ultrafast photonics applications. *Sci. Rep.* **2015**, *5*, 16372. [CrossRef]

71. Fu, B.; Hua, Y.; Xiao, X.; Zhu, H.; Sun, Z.; Yang, C. Broadband graphene saturable absorber for pulsed fiber lasers at 1, 1.5, and 2 μm. *IEEE J. Sel. Top. Quantum Electron.* **2014**, *20*, 411–415.

72. Song, Y.-W.; Jang, S.-Y.; Han, W.-S.; Bae, M.-K. Graphene mode-lockers for fiber lasers functioned with evanescent field interaction. *Appl. Phys. Lett.* **2010**, *96*, 051122. [CrossRef]

73. Yamashita, S. A tutorial on nonlinear photonic applications of carbon nanotube and graphene. *J. Lightwave Technol.* **2011**, *30*, 427–447. [CrossRef]

74. Zhang, H.; He, X.; Lin, W.; Wei, R.; Zhang, F.; Du, X.; Dong, G.; Qiu, J. Ultrafast saturable absorption in topological insulator Bi_2SeTe_2 nanosheets. *Opt. Express* **2015**, *23*, 13376–13383. [CrossRef] [PubMed]

75. Zhang, H.; Virally, S.; Bao, Q.; Kian Ping, L.; Massar, S.; Godbout, N.; Kockaert, P. Z-scan measurement of the nonlinear refractive index of graphene. *Opt. Lett.* **2012**, *37*, 1856–1858. [CrossRef] [PubMed]

76. Zheng, X.; Zhang, Y.; Chen, R.; Cheng, X.a.; Xu, Z.; Jiang, T. Z-scan measurement of the nonlinear refractive index of monolayer WS_2. *Opt. Express* **2015**, *23*, 15616–15623. [CrossRef] [PubMed]

77. Wang, G.; Wang, K.; Szydłowska, B.M.; Baker-Murray, A.A.; Wang, J.J.; Feng, Y.; Zhang, X.; Wang, J.; Blau, W.J. Ultrafast Nonlinear Optical Properties of a Graphene Saturable Mirror in the 2 μm Wavelength Region. *Laser Photonics Rev.* **2017**, *11*, 1700166. [CrossRef]

78. Chhowalla, M.; Shin, H.S.; Eda, G.; Li, L.-J.; Loh, K.P.; Zhang, H. The chemistry of two-dimensional layered transition metal dichalcogenide nanosheets. *Nat. Chem.* **2013**, *5*, 263–275. [CrossRef]

79. Ugeda, M.M.; Bradley, A.J.; Shi, S.-F.; da Jornada, F.H.; Zhang, Y.; Qiu, D.Y.; Ruan, W.; Mo, S.-K.; Hussain, Z.; Shen, Z.-X.; et al. Giant bandgap renormalization and excitonic effects in a monolayer transition metal dichalcogenide semiconductor. *Nat. Mater.* **2014**, *13*, 1091–1095. [CrossRef]

80. Lee, J.; Koo, J.; Lee, J.; Jhon, Y.M.; Lee, J.H. All-fiberized, femtosecond laser at 1912 nm using a bulk-like $MoSe_2$ saturable absorber. *Opt. Mater. Express* **2017**, *7*, 2968–2979. [CrossRef]

81. Cui, Y.; Lu, F.; Liu, X. Nonlinear Saturable and Polarization-induced Absorption of Rhenium Disulfide. *Sci. Rep.* **2017**, *7*, 40080. [CrossRef]

82. Wang, G.; Baker-Murray, A.A.; Blau, W.J. Saturable Absorption in 2D Nanomaterials and Related Photonic Devices. *Laser Photonics Rev.* **2019**, *13*, 1800282. [CrossRef]

83. Castellanos-Gomez, A.; Vicarelli, L.; Prada, E.; Island, J.O.; Narasimha-Acharya, K.; Blanter, S.I.; Groenendijk, D.J.; Buscema, M.; Steele, G.A.; Alvarez, J. Isolation and characterization of few-layer black phosphorus. *2D Mater.* **2014**, *1*, 025001. [CrossRef]

84. Xia, F.; Wang, H.; Jia, Y. Rediscovering black phosphorus as an anisotropic layered material for optoelectronics and electronics. *Nat. Commun.* **2014**, *5*, 4458. [CrossRef] [PubMed]

85. Sotor, J.; Sobon, G.; Macherzynski, W.; Paletko, P.; Abramski, K.M. Black phosphorus saturable absorber for ultrashort pulse generation. *Appl. Phys. Lett.* **2015**, *107*, 051108. [CrossRef]

86. Li, L.; Yu, Y.; Ye, G.J.; Ge, Q.; Ou, X.; Wu, H.; Feng, D.; Chen, X.H.; Zhang, Y. Black phosphorus field-effect transistors. *Nat. Nanotechnol.* **2014**, *9*, 372–377. [CrossRef] [PubMed]

87. Tran, V.; Soklaski, R.; Liang, Y.; Yang, L. Layer-controlled band gap and anisotropic excitons in few-layer black phosphorus. *Phys. Rev. B* **2014**, *89*, 235319. [CrossRef]

88. Liu, J.; Liu, J.; Guo, Z.; Zhang, H.; Ma, W.; Wang, J.; Su, L. Dual-wavelength Q-switched $Er:SrF_2$ laser with a black phosphorus absorber in the mid-infrared region. *Opt. Express* **2016**, *24*, 30289–30295. [CrossRef] [PubMed]

89. Wang, K.; Szydłowska, B.M.; Wang, G.; Zhang, X.; Wang, J.J.; Magan, J.J.; Zhang, L.; Coleman, J.N.; Wang, J.; Blau, W.J. Ultrafast Nonlinear Excitation Dynamics of Black Phosphorus Nanosheets from Visible to Mid-Infrared. *ACS Nano* **2016**, *10*, 6923–6932. [CrossRef] [PubMed]

90. Na, D.; Park, K.; Park, K.-H.; Song, Y.-W. Passivation of black phosphorus saturable absorbers for reliable pulse formation of fiber lasers. *Nanotechnology* **2017**, *28*, 475207. [CrossRef]

91. Moore, J.E. The birth of topological insulators. *Nature* **2010**, *464*, 194–198. [CrossRef]

92. Hasan, M.Z.; Kane, C.L. Colloquium: Topological insulators. *Rev. Mod. Phys.* **2010**, *82*, 3045–3067. [CrossRef]

93. Zhang, H.; Liu, C.-X.; Qi, X.-L.; Dai, X.; Fang, Z.; Zhang, S.-C. Topological insulators in Bi_2Se_3, Bi_2Te_3 and Sb_2Te_3 with a single Dirac cone on the surface. *Nat. Phys.* **2009**, *5*, 438–442. [CrossRef]

94. Naguib, M.; Come, J.; Dyatkin, B.; Presser, V.; Taberna, P.-L.; Simon, P.; Barsoum, M.W.; Gogotsi, Y. MXene: A promising transition metal carbide anode for lithium-ion batteries. *Electrochem. Commun.* **2012**, *16*, 61–64. [CrossRef]

95. Naguib, M.; Mochalin, V.N.; Barsoum, M.W.; Gogotsi, Y. 25th Anniversary Article: MXenes: A New Family of Two-Dimensional Materials. *Adv. Mater.* **2014**, *26*, 992–1005. [CrossRef] [PubMed]

96. Liu, M.-Y.; Huang, Y.; Chen, Q.-Y.; Li, Z.-Y.; Cao, C.; He, Y. Strain and electric field tunable electronic structure of buckled bismuthene. *RSC Adv.* **2017**, *7*, 39546–39555. [CrossRef]

97. Pumera, M.; Sofer, Z. 2D Monoelemental Arsenene, Antimonene, and Bismuthene: Beyond Black Phosphorus. *Adv. Mater.* **2017**, *29*, 1605299. [CrossRef]

98. Song, Y.; Liang, Z.; Jiang, X.; Chen, Y.; Li, Z.; Lu, L.; Ge, Y.; Wang, K.; Zheng, J.; Lu, S. Few-layer antimonene decorated microfiber: Ultra-short pulse generation and all-optical thresholding with enhanced long term stability. *2D Mater.* **2017**, *4*, 045010. [CrossRef]

99. Hong, S.; Lédée, F.; Park, J.; Song, S.; Lee, H.; Lee, Y.S.; Kim, B.; Yeom, D.I.; Deleporte, E.; Oh, K. Mode-Locking of All-Fiber Lasers Operating at Both Anomalous and Normal Dispersion Regimes in the C-and L-Bands Using Thin Film of 2D Perovskite Crystallites. *Laser Photonics Rev.* **2018**, *12*, 1800118. [CrossRef]

100. Liu, Z.; Mu, H.; Xiao, S.; Wang, R.; Wang, Z.; Wang, W.; Wang, Y.; Zhu, X.; Lu, K.; Zhang, H.; et al. Pulsed Lasers Employing Solution-Processed Plasmonic $Cu_{3-x}P$ Colloidal Nanocrystals. *Adv. Mater.* **2016**, *28*, 3535–3542. [CrossRef] [PubMed]

101. Lee, S.; Song, Y.-W. Graphene Self-Phase-Lockers Formed around a Cu Wire Hub for Ring Resonators Incorporated into 57.8 Gigahertz Fiber Pulsed Lasers. *ACS Nano* **2020**, *14*, 15944–15952. [CrossRef]

102. Wang, Y.; Mu, H.; Li, X.; Yuan, J.; Chen, J.; Xiao, S.; Bao, Q.; Gao, Y.; He, J. Observation of large nonlinear responses in a graphene-Bi_2Te_3 heterostructure at a telecommunication wavelength. *Appl. Phys. Lett.* **2016**, *108*, 221901. [CrossRef]

103. Frisenda, R.; Navarro-Moratalla, E.; Gant, P.; Pérez De Lara, D.; Jarillo-Herrero, P.; Gorbachev, R.V.; Castellanos-Gomez, A. Recent progress in the assembly of nanodevices and van der Waals heterostructures by deterministic placement of 2D materials. *Chem. Soc. Rev.* **2018**, *47*, 53–68. [CrossRef]

104. Wang, Z.; Mu, H.; Yuan, J.; Zhao, C.; Bao, Q.; Zhang, H. Graphene-Bi$_2$Te$_3$ heterostructure as broadband saturable absorber for ultra-short pulse generation in Er-doped and Yb-doped fiber lasers. *IEEE J. Sel. Top. Quantum Electron.* **2016**, *23*, 195–199. [CrossRef]

105. Chen, H.; Yin, J.; Yang, J.; Zhang, X.; Liu, M.; Jiang, Z.; Wang, J.; Sun, Z.; Guo, T.; Liu, W.; et al. Transition-metal dichalcogenides heterostructure saturable absorbers for ultrafast photonics. *Opt. Lett.* **2017**, *42*, 4279–4282. [CrossRef]

106. Ostojic, G.N.; Zaric, S.; Kono, J.; Strano, M.S.; Moore, V.C.; Hauge, R.H.; Smalley, R.E. Interband Recombination Dynamics in Resonantly Excited Single-Walled Carbon Nanotubes. *Phys. Rev. Lett.* **2004**, *92*, 117402. [CrossRef] [PubMed]

107. Lee, J.; Koo, J.; Jhon, Y.M.; Lee, J.H. A femtosecond pulse erbium fiber laser incorporating a saturable absorber based on bulk-structured Bi$_2$Te$_3$ topological insulator. *Opt. Express* **2014**, *22*, 6165–6173. [CrossRef] [PubMed]

108. Jung, M.; Lee, J.; Koo, J.; Park, J.; Song, Y.-W.; Lee, K.; Lee, S.; Lee, J.H. A femtosecond pulse fiber laser at 1935 nm using a bulk-structured Bi$_2$Te$_3$ topological insulator. *Opt. Express* **2014**, *22*, 7865–7874. [CrossRef] [PubMed]

109. Li, H.; Lu, G.; Wang, Y.; Yin, Z.; Cong, C.; He, Q.; Wang, L.; Ding, F.; Yu, T.; Zhang, H. Mechanical Exfoliation and Characterization of Single- and Few-Layer Nanosheets of WSe$_2$, TaS$_2$, and TaSe$_2$. *Small* **2013**, *9*, 1974–1981. [CrossRef]

110. Chen, Y.; Jiang, G.; Chen, S.; Guo, Z.; Yu, X.; Zhao, C.; Zhang, H.; Bao, Q.; Wen, S.; Tang, D.; et al. Mechanically exfoliated black phosphorus as a new saturable absorber for both Q-switching and Mode-locking laser operation. *Opt. Express* **2015**, *23*, 12823–12833. [CrossRef]

111. Wang, Z.; Jia, H.; Zheng, X.; Yang, R.; Wang, Z.; Ye, G.; Chen, X.; Shan, J.; Feng, P.X.-L. Black phosphorus nanoelectromechanical resonators vibrating at very high frequencies. *Nanoscale* **2015**, *7*, 877–884. [CrossRef] [PubMed]

112. Nicolosi, V.; Chhowalla, M.; Kanatzidis, M.G.; Strano, M.S.; Coleman, J.N. Liquid exfoliation of layered materials. *Science* **2013**, *340*, 6139. [CrossRef]

113. Aiub, E.J.; Steinberg, D.; de Souza, E.A.T.; Saito, L.A. 200-fs mode-locked Erbium-doped fiber laser by using mechanically exfoliated MoS$_2$ saturable absorber onto D-shaped optical fiber. *Opt. Express* **2017**, *25*, 10546–10552. [CrossRef]

114. Luo, Z.-C.; Liu, M.; Guo, Z.-N.; Jiang, X.-F.; Luo, A.-P.; Zhao, C.-J.; Yu, X.-F.; Xu, W.-C.; Zhang, H. Microfiber-based few-layer black phosphorus saturable absorber for ultra-fast fiber laser. *Opt. Express* **2015**, *23*, 20030–20039. [CrossRef]

115. Li, X.; Cai, W.; An, J.; Kim, S.; Nah, J.; Yang, D.; Piner, R.; Velamakanni, A.; Jung, I.; Tutuc, E. Large-area synthesis of high-quality and uniform graphene films on copper foils. *Science* **2009**, *324*, 1312–1314. [CrossRef]

116. Liu, W.; Liu, M.; Yin, J.; Chen, H.; Lu, W.; Fang, S.; Teng, H.; Lei, M.; Yan, P.; Wei, Z. Tungsten diselenide for all-fiber lasers with the chemical vapor deposition method. *Nanoscale* **2018**, *10*, 7971–7977. [CrossRef]

117. Yin, J.; Li, J.; Chen, H.; Wang, J.; Yan, P.; Liu, M.; Liu, W.; Lu, W.; Xu, Z.; Zhang, W. Large-area highly crystalline WSe$_2$ atomic layers for ultrafast pulsed lasers. *Opt. Express* **2017**, *25*, 30020–30031. [CrossRef] [PubMed]

118. Uddin, S.; Song, Y.-W. Atomic Carbon Spraying: Direct Growth of Graphene on Customized 3D Surfaces of Ultrafast Optical Devices. *Adv. Opt. Mater.* **2020**, *8*, 1902091. [CrossRef]

119. Kim, W.-J.; Debnath, P.C.; Lee, J.; Lee, J.H.; Lim, D.-S.; Song, Y.-W. Transfer-free synthesis of multilayer graphene using a single-step process in an evaporator and formation confirmation by laser mode-locking. *Nanotechnology* **2013**, *24*, 365603. [CrossRef] [PubMed]

120. Debnath, P.C.; Uddin, S.; Song, Y.-W. Ultrafast all-optical switching incorporating in situ graphene grown along an optical fiber by the evanescent field of a laser. *ACS Photonics* **2018**, *5*, 445–455. [CrossRef]

121. Reina, A.; Son, H.; Jiao, L.; Fan, B.; Dresselhaus, M.S.; Liu, Z.; Kong, J. Transferring and Identification of Single- and Few-Layer Graphene on Arbitrary Substrates. *J. Phys. Chem. C* **2008**, *112*, 17741–17744. [CrossRef]

122. Song, Y.-W.; Yamashita, S.; Goh, C.S.; Set, S.Y. Carbon nanotube mode lockers with enhanced nonlinearity via evanescent field interaction in D-shaped fibers. *Opt. Lett.* **2007**, *32*, 148–150. [CrossRef]

123. Song, Y.-W.; Morimune, K.; Set, S.Y.; Yamashita, S. Polarization insensitive all-fiber mode-lockers functioned by carbon nanotubes deposited onto tapered fibers. *Appl. Phys. Lett.* **2007**, *90*, 021101. [CrossRef]

124. Zhao, J.; Ruan, S.; Yan, P.; Zhang, H.; Yu, Y.; Wei, H.; Luo, J. Cladding-filled graphene in a photonic crystal fiber as a saturable absorber and its first application for ultrafast all-fiber laser. *Opt. Eng.* **2013**, *52*, 106105. [CrossRef]

125. Lin, Y.-H.; Yang, C.-Y.; Liou, J.-H.; Yu, C.-P.; Lin, G.-R. Using graphene nano-particle embedded in photonic crystal fiber for evanescent wave mode-locking of fiber laser. *Opt. Express* **2013**, *21*, 16763–16776. [CrossRef]

126. Woodward, R.I.; Howe, R.C.T.; Runcorn, T.H.; Hu, G.; Torrisi, F.; Kelleher, E.J.R.; Hasan, T. Wideband saturable absorption in few-layer molybdenum diselenide (MoSe$_2$) for Q-switching Yb-, Er- and Tm-doped fiber lasers. *Opt. Express* **2015**, *23*, 20051–20061. [CrossRef]

127. Liu, X.M.; Yang, H.R.; Cui, Y.D.; Chen, G.W.; Yang, Y.; Wu, X.Q.; Yao, X.K.; Han, D.D.; Han, X.X.; Zeng, C.; et al. Graphene-clad microfibre saturable absorber for ultrafast fibre lasers. *Sci. Rep.* **2016**, *6*, 26024. [CrossRef]

128. Park, N.H.; Jeong, H.; Choi, S.Y.; Kim, M.H.; Rotermund, F.; Yeom, D.I. Monolayer graphene saturable absorbers with strongly enhanced evanescent-field interaction for ultrafast fiber laser mode-locking. *Opt. Express* **2015**, *23*, 19806–19812. [CrossRef]

129. Purdie, D.G.; Popa, D.; Wittwer, V.J.; Jiang, Z.; Bonacchini, G.; Torrisi, F.; Milana, S.; Lidorikis, E.; Ferrari, A.C. Few-cycle pulses from a graphene mode-locked all-fiber laser. *Appl. Phys. Lett.* **2015**, *106*, 253101. [CrossRef]

130. Choi, S.Y.; Jeong, H.; Hong, B.H.; Rotermund, F.; Yeom, D.-I. All-fiber dissipative soliton laser with 10.2 nJ pulse energy using an evanescent field interaction with graphene saturable absorber. *Laser Phys. Lett.* **2013**, *11*, 015101. [CrossRef]

131. Sotor, J.; Sobon, G.; Abramski, K.M. Sub-130 fs mode-locked Er-doped fiber laser based on topological insulator. *Opt. Express* **2014**, *22*, 13244–13249. [CrossRef]

132. Gao, L.; Zhu, T.; Huang, W.; Luo, Z. Stable, Ultrafast Pulse Mode-Locked by Topological Insulator Bi_2Se_3 Nanosheets Interacting With Photonic Crystal Fiber: From Anomalous Dispersion to Normal Dispersion. *IEEE Photonics J.* **2015**, *7*, 1–8.

133. Liu, W.; Pang, L.; Han, H.; Liu, M.; Lei, M.; Fang, S.; Teng, H.; Wei, Z. Tungsten disulfide saturable absorbers for 67 fs mode-locked erbium-doped fiber lasers. *Opt. Express* **2017**, *25*, 2950–2959. [CrossRef]

134. Wang, J.; Jiang, Z.; Chen, H.; Li, J.; Yin, J.; Wang, J.; He, T.; Yan, P.; Ruan, S. High energy soliton pulse generation by a magnetron-sputtering-deposition-grown $MoTe_2$ saturable absorber. *Photonics Res.* **2018**, *6*, 535–541. [CrossRef]

135. Jin, X.; Hu, G.; Zhang, M.; Hu, Y.; Albrow-Owen, T.; Howe, R.C.T.; Wu, T.C.; Wu, Q.; Zheng, Z.; Hasan, T. 102 fs pulse generation from a long-term stable, inkjet-printed black phosphorus-mode-locked fiber laser. *Opt. Express* **2018**, *26*, 12506–12513. [CrossRef]

136. Kim, D.; Park, N.H.; Lee, H.; Lee, J.; Yeom, D.-I.; Kim, J. Graphene-based saturable absorber and mode-locked laser behaviors under gamma-ray radiation. *Photonics Res.* **2019**, *7*, 742–747. [CrossRef]

137. Bao, Q.; Zhang, H.; Ni, Z.; Wang, Y.; Polavarapu, L.; Shen, Z.; Xu, Q.-H.; Tang, D.; Loh, K.P. Monolayer graphene as a saturable absorber in a mode-locked laser. *Nano Res.* **2011**, *4*, 297–307. [CrossRef]

138. Los, J.H.; Zakharchenko, K.V.; Katsnelson, M.I.; Fasolino, A. Melting temperature of graphene. *Phys. Rev. B* **2015**, *91*, 045415. [CrossRef]

139. Huang, P.L.; Lin, S.C.; Yeh, C.Y.; Kuo, H.H.; Huang, S.H.; Lin, G.R.; Li, L.J.; Su, C.Y.; Cheng, W.H. Stable mode-locked fiber laser based on CVD fabricated graphene saturable absorber. *Opt. Express* **2012**, *20*, 2460–2465. [CrossRef]

140. Martinez, A.; Fuse, K.; Yamashita, S. Mechanical exfoliation of graphene for the passive mode-locking of fiber lasers. *Appl. Phys. Lett.* **2011**, *99*, 121107. [CrossRef]

141. Sobon, G.; Sotor, J.; Abramski, K. All-polarization maintaining femtosecond Er-doped fiber laser mode-locked by graphene saturable absorber. *Laser Phys. Lett.* **2012**, *9*, 581. [CrossRef]

142. Sotor, J.; Pasternak, I.; Krajewska, A.; Strupinski, W.; Sobon, G. Sub-90 fs a stretched-pulse mode-locked fiber laser based on a graphene saturable absorber. *Opt. Express* **2015**, *23*, 27503–27508. [CrossRef]

143. Sun, Z.; Popa, D.; Hasan, T.; Torrisi, F.; Wang, F.; Kelleher, E.J.; Travers, J.C.; Nicolosi, V.; Ferrari, A.C. A stable, wideband tunable, near transform-limited, graphene-mode-locked, ultrafast laser. *Nano Res.* **2010**, *3*, 653–660. [CrossRef]

144. Zhou, Y.; Lin, J.; Zhang, X.; Xu, L.; Gu, C.; Sun, B.; Wang, A.; Zhan, Q. Self-starting passively mode-locked all fiber laser based on carbon nanotubes with radially polarized emission. *Photonics Res.* **2016**, *4*, 327–330. [CrossRef]

145. Wu, K.; Li, X.; Wang, Y.; Wang, Q.J.; Shum, P.P.; Chen, J. Towards low timing phase noise operation in fiber lasers mode locked by graphene oxide and carbon nanotubes at 1.5 μm. *Opt. Express* **2015**, *23*, 501–511. [CrossRef]

146. Lau, K.Y.; Ker, P.J.; Abas, A.F.; Alresheedi, M.T.; Mahdi, M.A. Mode-locked fiber laser in the C-band region for dual-wavelength ultrashort pulses emission using a carbon nanotube saturable absorber. *Chin. Opt. Lett.* **2019**, *17*, 051401. [CrossRef]

147. Yemineni, S.R.; Arokiaswami, A.; Shum, P. All-fiber femtosecond laser pulse generation at 1.55 μm and 2 μm using a common carbon-nanotube based saturable absorber. In Proceedings of the 2017 IEEE Conference on Lasers and Electro-Optics Pacific Rim (CLEO-PR), Singapore, 31 July–4 August 2017; pp. 1–2.

148. Lazarev, V.; Krylov, A.; Dvoretskiy, D.; Sazonkin, S.; Pnev, A.; Leonov, S.; Shelestov, D.; Tarabrin, M.; Karasik, V.; Kireev, A. Stable similariton generation in an all-fiber hybrid mode-locked ring laser for frequency metrology. *IEEE Trans. Ultrason. Ferroelectr. Freq. Control.* **2016**, *63*, 1028–1033. [CrossRef]

149. Chen, Y.; Chen, S.; Liu, J.; Gao, Y.; Zhang, W. Sub-300 femtosecond soliton tunable fiber laser with all-anomalous dispersion passively mode locked by black phosphorus. *Opt. Express* **2016**, *24*, 13316–13324. [CrossRef]

150. Jin, X.; Hu, G.; Zhang, M.; Hu, Y.; Albrow-Owen, T.; Howe, R.; Wu, T.; Zhu, X.; Zheng, Z.; Hasan, T. Long term stable black phosphorus saturable absorber for mode-locked fiber laser. In Proceedings of the 2017 Conference on Lasers and Electro-Optics (CLEO): Science and Innovations, Optical Society of America, San Jose, CA, USA, 14–19 May 2017; p. SW4K. 1.

151. Liu, M.; Zhao, N.; Liu, H.; Zheng, X.-W.; Luo, A.-P.; Luo, Z.-C.; Xu, W.-C.; Zhao, C.-J.; Zhang, H.; Wen, S.-C. Dual-wavelength harmonically mode-locked fiber laser with topological insulator saturable absorber. *IEEE Photonics Technol. Lett.* **2014**, *26*, 983–986.

152. Li, K.; Song, Y.; Yu, Z.; Xu, R.; Dou, Z.; Tian, J. L-band femtosecond fibre laser based on Bi_2Se_3 topological insulator. *Laser Phys. Lett.* **2015**, *12*, 105103. [CrossRef]

153. Haris, H.; Arof, H.; Muhammad, A.R.; Anyi, C.L.; Tan, S.J.; Kasim, N.; Harun, S.W. Passively Q-switched and mode-locked Erbium-doped fiber laser with topological insulator Bismuth Selenide (Bi_2Se_3) as saturable absorber at C-band region. *Opt. Fiber Technol.* **2019**, *48*, 117–122. [CrossRef]

154. Guo, Q.; Pan, J.; Liu, Y.; Si, H.; Lu, Z.; Han, X.; Gao, J.; Zuo, Z.; Zhang, H.; Jiang, S. Output energy enhancement in a mode-locked Er-doped fiber laser using CVD-Bi_2Se_3 as a saturable absorber. *Opt. Express* **2019**, *27*, 24670–24681. [CrossRef]

155. Lin, Y.-H.; Yang, C.-Y.; Lin, S.-F.; Tseng, W.-H.; Bao, Q.; Wu, C.-I.; Lin, G.-R. Soliton compression of the erbium-doped fiber laser weakly started mode-locking by nanoscale p-type Bi_2Te_3 topological insulator particles. *Laser Phys. Lett.* **2014**, *11*, 055107. [CrossRef]

156. Lee, J.; Koo, J.; Lee, J.H. A pulse-width-tunable, mode-locked fiber laser based on dissipative soliton resonance using a bulk-structured Bi_2Te_3 topological insulator. *Opt. Eng.* **2016**, *55*, 081309. [CrossRef]

157. Luo, Z.-C.; Liu, M.; Liu, H.; Zheng, X.-W.; Luo, A.-P.; Zhao, C.-J.; Zhang, H.; Wen, S.-C.; Xu, W.-C. 2 GHz passively harmonic mode-locked fiber laser by a microfiber-based topological insulator saturable absorber. *Opt. Lett.* **2013**, *38*, 5212–5215. [CrossRef]

158. Wei, Q.; Niu, K.; Han, X.; Zhang, H.; Zhang, C.; Yang, C.; Man, B. Large energy pulses generation in a mode-locked Er-doped fiber laser based on CVD-grown Bi_2Te_3 saturable absorber. *Opt. Mater. Express* **2019**, *9*, 3535–3545. [CrossRef]
159. Yan, P.; Lin, R.; Ruan, S.; Liu, A.; Chen, H.; Zheng, Y.; Chen, S.; Guo, C.; Hu, J. A practical topological insulator saturable absorber for mode-locked fiber laser. *Sci. Rep.* **2015**, *5*, 1–5. [CrossRef]
160. Liu, C.; Li, H.; Deng, G.; Lan, C.; Li, C.; Liu, Y. Femtosecond Er-Doped Fiber Laser Using a Graphene/MoS_2 Heterostructure Saturable Absorber. In Proceedings of the 2016 Asia Communications and Photonics Conference (ACP), Wuhan, China, 2–5 November 2016; pp. 1–3.
161. Ahmed, M.; Latiff, A.; Arof, H.; Harun, S.W. Mode-locking pulse generation with MoS_2–PVA saturable absorber in both anomalous and ultra-long normal dispersion regimes. *Appl. Opt.* **2016**, *55*, 4247–4252. [CrossRef] [PubMed]
162. Khazaeizhad, R.; Kassani, S.H.; Jeong, H.; Yeom, D.I.; Oh, K. Mode-locking of Er-doped fiber laser using a multilayer MoS_2 thin film as a saturable absorber in both anomalous and normal dispersion regimes. *Opt. Express* **2014**, *22*, 23732–23742. [CrossRef] [PubMed]
163. Liu, M.; Zheng, X.-W.; Qi, Y.-L.; Liu, H.; Luo, A.-P.; Luo, Z.-C.; Xu, W.-C.; Zhao, C.-J.; Zhang, H. Microfiber-based few-layer MoS_2 saturable absorber for 2.5 GHz passively harmonic mode-locked fiber laser. *Opt. Express* **2014**, *22*, 22841–22846. [CrossRef]
164. Liu, H.; Luo, A.-P.; Wang, F.-Z.; Tang, R.; Liu, M.; Luo, Z.-C.; Xu, W.-C.; Zhao, C.-J.; Zhang, H. Femtosecond pulse erbium-doped fiber laser by a few-layer MoS_2 saturable absorber. *Opt. Lett.* **2014**, *39*, 4591–4594. [CrossRef]
165. Wu, K.; Zhang, X.; Wang, J.; Chen, J. 463-MHz fundamental mode-locked fiber laser based on few-layer MoS_2 saturable absorber. *Opt. Lett.* **2015**, *40*, 1374–1377. [CrossRef]
166. Wang, Y.; Mao, D.; Gan, X.; Han, L.; Ma, C.; Xi, T.; Zhang, Y.; Shang, W.; Hua, S.; Zhao, J. Harmonic mode locking of bound-state solitons fiber laser based on MoS_2 saturable absorber. *Opt. Express* **2015**, *23*, 205–210. [CrossRef]
167. Koo, J.; Park, J.; Lee, J.; Jhon, Y.M.; Lee, J.H. Femtosecond harmonic mode-locking of a fiber laser at 3.27 GHz using a bulk-like, $MoSe_2$-based saturable absorber. *Opt. Express* **2016**, *24*, 10575–10589. [CrossRef] [PubMed]
168. Ahmad, H.B.; Aidit, S.N.; Hassan, N.A.; Ismail, M.F.; Tiu, Z.C. Generation of mode-locked erbium-doped fiber laser using $MoSe_2$ as saturable absorber. *Opt. Eng.* **2016**, *55*, 076115. [CrossRef]
169. Liu, W.; Liu, M.; OuYang, Y.; Hou, H.; Lei, M.; Wei, Z. CVD-grown $MoSe_2$ with high modulation depth for ultrafast mode-locked erbium-doped fiber laser. *Nanotechnology* **2018**, *29*, 394002. [CrossRef] [PubMed]
170. Zhang, R.-L.; Wang, J.; Zhang, X.-Y.; Lin, J.-T.; Li, X.; Kuan, P.-W.; Zhou, Y.; Liao, M.-S.; Gao, W.-Q. Mode-locked fiber laser with $MoSe_2$ saturable absorber based on evanescent field. *Chin. Phys. B* **2019**, *28*, 014207. [CrossRef]
171. Liu, M.; Liu, W.; Wei, Z. $MoTe_2$ saturable absorber with high modulation depth for erbium-doped fiber laser. *J. Lightwave Technol.* **2019**, *37*, 3100–3105. [CrossRef]
172. Han, X. 2D $MoTe_2$ film as a saturable absorber for a wavelength-tunable ultrafast fiber laser. *Appl. Opt.* **2019**, *58*, 8390–8395. [CrossRef]
173. Khazaeinezhad, R.; Kassani, S.H.; Jeong, H.; Yeom, D.-I.; Oh, K. Femtosecond soliton pulse generation using evanescent field interaction through Tungsten disulfide (WS_2) film. *J. Lightwave Technol.* **2015**, *33*, 3550–3557. [CrossRef]
174. Wu, K.; Zhang, X.; Wang, J.; Li, X.; Chen, J. WS_2 as a saturable absorber for ultrafast photonic applications of mode-locked and Q-switched lasers. *Opt. Express* **2015**, *23*, 11453–11461. [CrossRef]
175. Mao, D.; Zhang, S.; Wang, Y.; Gan, X.; Zhang, W.; Mei, T.; Wang, Y.; Wang, Y.; Zeng, H.; Zhao, J. WS_2 saturable absorber for dissipative soliton mode locking at 1.06 and 1.55 μm. *Opt. Express* **2015**, *23*, 27509–27519. [CrossRef]
176. Yan, P.; Liu, A.; Chen, Y.; Chen, H.; Ruan, S.; Guo, C.; Chen, S.; Li, I.L.; Yang, H.; Hu, J. Microfiber-based WS_2-film saturable absorber for ultra-fast photonics. *Opt. Mater. Express* **2015**, *5*, 479–489. [CrossRef]
177. Yan, P.; Liu, A.; Chen, Y.; Wang, J.; Ruan, S.; Chen, H.; Ding, J. Passively mode-locked fiber laser by a cell-type WS_2 nanosheets saturable absorber. *Sci. Rep.* **2015**, *5*, 1–7. [CrossRef] [PubMed]
178. Koo, J.; Jhon, Y.I.; Park, J.; Lee, J.; Jhon, Y.M.; Lee, J.H. Near-Infrared saturable absorption of defective bulk-structured WTe_2 for femtosecond laser mode-locking. *Adv. Funct. Mater.* **2016**, *26*, 7454–7461. [CrossRef]
179. Wu, Q.; Zhang, M.; Jin, X.; Chen, S.; Jiang, Q.; Jiang, X.; Zheng, Z.; Zhang, H. 104 fs mode-locked fiber laser with a MXene-based saturable absorber. In Proceedings of the 2019 Conference on Lasers and Electro-Optics (CLEO): Applications and Technology, Optical Society of America, San Jose, CA, USA, 5–10 May 2019; p. JW2A. 86.
180. Jiang, X.; Liu, S.; Liang, W.; Luo, S.; He, Z.; Ge, Y.; Wang, H.; Cao, R.; Zhang, F.; Wen, Q. Broadband nonlinear photonics in few-layer MXene $Ti_3C_2T_x$ (T= F, O, or OH). *Laser Photonics Rev.* **2018**, *12*, 1700229. [CrossRef]
181. Yi, J.; Du, L.; Li, J.; Yang, L.; Hu, L.; Huang, S.; Dong, Y.; Miao, L.; Wen, S.; Mochalin, V.N. Unleashing the potential of Ti_2CT_x MXene as a pulse modulator for mid-infrared fiber lasers. *2D Mater.* **2019**, *6*, 045038. [CrossRef]
182. Du, J.; Zhang, M.; Zhu, X.; Hu, G.; Zhao, X.; Zheng, Z.; Zhang, H. Microfiber-based few-layer black phosphorus quantum dots saturable absorber for mode-locked fiber laser. In Proceedings of the 2016 IEEE Conference on Lasers and Electro-Optics (CLEO), San Jose, CA, USA, 5–10 June 2016; pp. 1–2.
183. Set, S.Y.; Yaguchi, H.; Tanaka, Y.; Jablonski, M.; Sakakibara, Y.; Rozhin, A.; Tokumoto, M.; Kataura, H.; Achiba, Y.; Kikuchi, K. Mode-locked Fiber Lasers based on a Saturable Absorber Incorporating Carbon Nanotubes. In Proceedings of the Optical Fiber Communication Conference, Atlanta, Georgia, 23 March 2003; p. PD44.
184. Martinez, A.; Al Araimi, M.; Dmitriev, A.; Lutsyk, P.; Li, S.; Mou, C.; Rozhin, A.; Sumetsky, M.; Turitsyn, S. Low-loss saturable absorbers based on tapered fibers embedded in carbon nanotube/polymer composites. *APL Photonics* **2017**, *2*, 126103. [CrossRef]

185. Martinez, A.; Zhou, K.; Bennion, I.; Yamashita, S. In-fiber microchannel device filled with a carbon nanotube dispersion for passive mode-lock lasing. *Opt. Express* **2008**, *16*, 15425–15430. [CrossRef]
186. Martinez, A.; Zhou, K.; Bennion, I.; Yamashita, S. Passive mode-locked lasing by injecting a carbon nanotube-solution in the core of an optical fiber. *Opt. Express* **2010**, *18*, 11008–11014. [CrossRef]
187. Li, Y.; Gao, L.; Huang, W.; Gao, C.; Liu, M.; Zhu, T. All-fiber mode-locked laser via short single-wall carbon nanotubes interacting with evanescent wave in photonic crystal fiber. *Opt. Express* **2016**, *24*, 23450–23458. [CrossRef]
188. Choi, S.Y.; Rotermund, F.; Jung, H.; Oh, K.; Yeom, D.I. Femtosecond mode-locked fiber laser employing a hollow optical fiber filled with carbon nanotube dispersion as saturable absorber. *Opt. Express* **2009**, *17*, 21788–21793. [CrossRef]
189. Hisyam, M.B.; Rusdi, M.F.M.; Latiff, A.A.; Harun, S.W. Generation of mode-locked ytterbium doped fiber ring laser using few-layer black phosphorus as a saturable absorber. *IEEE J. Sel. Top. Quantum Electron.* **2016**, *23*, 39–43. [CrossRef]
190. Pawliszewska, M.; Ge, Y.; Li, Z.; Zhang, H.; Sotor, J. Fundamental and harmonic mode-locking at 2.1 mm with black phosphorus saturable absorber. *Opt. Express* **2017**, *25*, 16916–16921. [CrossRef]
191. Lu, L.; Liang, Z.; Wu, L.; Chen, Y.; Song, Y.; Dhanabalan, S.C.; Ponraj, J.S.; Dong, B.; Xiang, Y.; Xing, F.; et al. Few-layer Bismuthene: Sonochemical Exfoliation, Nonlinear Optics and Applications for Ultrafast Photonics with Enhanced Stability. *Laser Photonics Rev.* **2018**, *12*, 1700221. [CrossRef]
192. Island, J.O.; Steele, G.A.; van der Zant, H.S.; Castellanos-Gomez, A. Environmental instability of few-layer black phosphorus. *2D Mater.* **2015**, *2*, 011002. [CrossRef]
193. Zhang, Y.; Yu, H.; Zhang, R.; Zhao, G.; Zhang, H.; Chen, Y.; Mei, L.; Tonelli, M.; Wang, J. Broadband atomic-layer MoS_2 optical modulators for ultrafast pulse generations in the visible range. *Opt. Lett.* **2017**, *42*, 547–550. [CrossRef] [PubMed]
194. Wang, S.; Yu, H.; Zhang, H.; Wang, A.; Zhao, M.; Chen, Y.; Mei, L.; Wang, J. Broadband Few-Layer MoS_2 Saturable Absorbers. *Adv. Mater.* **2014**, *26*, 3538–3544. [CrossRef] [PubMed]
195. Khazaeinezhad, R.; Kassani, S.H.; Jeong, H.; Park, K.J.; Kim, B.Y.; Yeom, D.-I.; Oh, K. Ultrafast pulsed all-fiber laser based on tapered fiber enclosed by few-layer WS_2 nanosheets. *IEEE Photonics Technol. Lett.* **2015**, *27*, 1581–1584. [CrossRef]
196. Wang, K.; Wang, J.; Fan, J.; Lotya, M.; O'Neill, A.; Fox, D.; Feng, Y.; Zhang, X.; Jiang, B.; Zhao, Q.; et al. Ultrafast Saturable Absorption of Two-Dimensional MoS_2 Nanosheets. *ACS Nano* **2013**, *7*, 9260–9267. [CrossRef]
197. Zhang, H.; Lu, S.B.; Zheng, J.; Du, J.; Wen, S.C.; Tang, D.Y.; Loh, K.P. Molybdenum disulfide (MoS_2) as a broadband saturable absorber for ultra-fast photonics. *Opt. Express* **2014**, *22*, 7249–7260. [CrossRef]
198. Manzeli, S.; Ovchinnikov, D.; Pasquier, D.; Yazyev, O.V.; Kis, A. 2D transition metal dichalcogenides. *Nat. Rev. Mater.* **2017**, *2*, 17033. [CrossRef]
199. Ouyang, Q.; Yu, H.; Zhang, K.; Chen, Y. Saturable absorption and the changeover from saturable absorption to reverse saturable absorption of MoS_2 nanoflake array films. *J. Mater. Chem. C* **2014**, *2*, 6319–6325. [CrossRef]
200. Xia, H.; Li, H.; Lan, C.; Li, C.; Zhang, X.; Zhang, S.; Liu, Y. Ultrafast erbium-doped fiber laser mode-locked by a CVD-grown molybdenum disulfide (MoS_2) saturable absorber. *Opt. Express* **2014**, *22*, 17341–17348. [CrossRef] [PubMed]
201. Zhao, C.; Zhang, H.; Qi, X.; Chen, Y.; Wang, Z.; Wen, S.; Tang, D. Ultra-short pulse generation by a topological insulator based saturable absorber. *Appl. Phys. Lett.* **2012**, *101*, 211106. [CrossRef]
202. Chen, S.; Zhao, C.; Li, Y.; Huang, H.; Lu, S.; Zhang, H.; Wen, S. Broadband optical and microwave nonlinear response in topological insulator. *Opt. Mater. Express* **2014**, *4*, 587–596. [CrossRef]
203. Yu, H.; Zhang, H.; Wang, Y.; Zhao, C.; Wang, B.; Wen, S.; Zhang, H.; Wang, J. Topological insulator as an optical modulator for pulsed solid-state lasers. *Laser Photonics Rev.* **2013**, *7*, L77–L83. [CrossRef]
204. Dou, Z.; Song, Y.; Tian, J.; Liu, J.; Yu, Z.; Fang, X. Mode-locked ytterbium-doped fiber laser based on topological insulator: Bi_2Se_3. *Opt. Express* **2014**, *22*, 24055–24061. [CrossRef] [PubMed]
205. Zhao, C.; Zou, Y.; Chen, Y.; Wang, Z.; Lu, S.; Zhang, H.; Wen, S.; Tang, D. Wavelength-tunable picosecond soliton fiber laser with Topological Insulator: Bi_2Se_3 as a mode locker. *Opt. Express* **2012**, *20*, 27888–27895. [CrossRef] [PubMed]
206. Li, W.; Chen, B.; Meng, C.; Fang, W.; Xiao, Y.; Li, X.; Hu, Z.; Xu, Y.; Tong, L.; Wang, H.; et al. Ultrafast All-Optical Graphene Modulator. *Nano Lett.* **2014**, *14*, 955–959. [CrossRef]
207. Sheng, Q.-W.; Feng, M.; Xin, W.; Guo, H.; Han, T.-Y.; Li, Y.-G.; Liu, Y.-G.; Gao, F.; Song, F.; Liu, Z.-B.; et al. Tunable graphene saturable absorber with cross absorption modulation for mode-locking in fiber laser. *Appl. Phys. Lett.* **2014**, *105*, 041901. [CrossRef]
208. Girard, S.; Morana, A.; Ladaci, A.; Robin, T.; Mescia, L.; Bonnefois, J.-J.; Boutillier, M.; Mekki, J.; Paveau, A.; Cadier, B. Recent advances in radiation-hardened fiber-based technologies for space applications. *J. Opt.* **2018**, *20*, 093001. [CrossRef]
209. Fortescue, P.; Swinerd, G.; Stark, J. *Spacecraft Systems Engineering*, 4th ed.; John Wiley & Sons: Hoboken, NJ, USA, 2011; p. 399.

 *sensors*

Article

Characteristic Test Analysis of Graphene Plus Optical Microfiber Coupler Combined Device and Its Application in Fiber Lasers

Yang Yu [1,2,*], Hao Chen [3], Zhenfu Zhang [1], Dingbo Chen [1], Jianfei Wang [2], Zhengtong Wei [4], Junbo Yang [1] and Peiguang Yan [3]

[1] Department of Physics, College of Liberal Arts and Sciences, National University of Defense Technology, Changsha 410073, China; zhangzhenfu198206@163.com (Z.Z.); chendingbo15@nudt.edu.cn (D.C.); yangjunbo@nudt.edu.cn (J.Y.)
[2] Deep Sea Technology Laboratory, College of Meteorology and Oceanology, National University of Defense Technology, Changsha 410073, China; wangjianfei09@nudt.edu.cn
[3] Shenzhen Key Lab of Laser Technology, College of Optoelectronic Engineering, Shenzhen University, Shenzhen 518060, China; hchenhao@szu.edu.cn (H.C.); yanpg@szu.edu.cn (P.Y.)
[4] Department of Basic Education, Information Engineering University, Zhengzhou 450000, China; weizhengtong1987@126.com
* Correspondence: yuyang08a@nudt.edu.cn; Tel.: +86-186-7001-3227

Received: 20 February 2020; Accepted: 12 March 2020; Published: 16 March 2020

Abstract: In this study, a graphene and optical microfiber coupler (OMC) integrated device (GOMC) was proposed and fabricated. After its characteristic analysis and testing, it was applied to the development of adjustable multi-wavelength fiber lasers. By integrating the OMC with graphene, the polarization dependence of OMC was enhanced. Meanwhile, the novel GOMC was given the capabilities of filtering, coupling, beam splitting, and polarization correlation. When the GOMC was integrated as a filter and beam splitter into the ring cavity of the fiber laser, the proposed GOMC-based fiber laser could achieve single-wavelength and multi-wavelength regulated output. The laser had a 3 dB linewidth of less than 30 pm, a signal-to-noise ratio of approximately 40 dB, and an output power fluctuation of less than 1 dB. The GOMC could also be used for the development of functional devices, such as adjustable mode lockers and mode coupling selectors, which provide an excellent experimental platform for new fiber lasers and the research of multi-dimensional light-field manipulation.

Keywords: optical microfiber coupler; graphene; fiber laser; filter

---

1. Introduction

In recent years, owing to the advantages of easy integration with existing optical systems and the high sensitivity to external environmental changes, optical microfiber couplers (OMCs) have been widely used in sensing fields and other functional devices (especially by the fusing and tapering of two conventional fibers) [1–8]. Moreover, the OMC is a wavelength-dependent device that can be used as a mode-selection filter for fiber lasers [5]. Compared with the traditional filters, OMC-based fiber filters possess the advantages of easy fabrication, low cost, compact configuration, multi-transmission ports, and especially flexible structure design to realize different filter functions. In addition, due to the high heat transfer efficiency of the waist region, the optical transmission characteristics of the OMC are extremely sensitive to temperature changes [6]. Benefiting from the above-mentioned advantages, we developed an all-optical modulator and an all-optical tunable filter using OMCs [7,8]. Recently, it has been reported that mode-selective coupled OMCs can also be used in a vector fiber laser system [9]. Since the OMC has a strong function expansion, this leads to a great application value in the development of new fiber lasers. However, the transaction always has a potential disadvantage. When the OMC is integrated into the resonator of the fiber laser as a functional device, such as a filter or a beam splitter, if the package and temperature control are not performed effectively, the stability of the laser will be affected by the environmental sensitivity of the functional device. Therefore, further studies on functional devices based on OMCs are essential for different applications.

In the past ten years, two-dimensional (2D) materials, especially graphene, has attracted attention in the applications of electronics, photonics, and optoelectronics, thanks to their tunable Fermi level, saturable absorption, broad-spectrum band filling effect, ultrafast electron migrating rate, higher third-order nonlinear coefficients, and other excellent photoelectric characteristics [10,11]. In addition, 2D materials also have an ultra-high heat transfer efficiency such that they are also ideal materials for temperature control [12]. Due to the large evanescent field transmission properties of optical microfibers (OMs), its integration with 2D materials can effectively enhance the interaction between light and materials, which have been developed into fiber optic functional devices, such as polarizers, modulators, and saturated absorbers (mode lockers) [13–16]. On one hand, these functional devices are easy to integrate with existing fiber optic systems, and on the other hand, they make full use of the unique optical characteristics of two-dimensional materials to effectively implement multi-dimensional regulation of the light-field. Moreover, the cross-fusion of micro-nano photonics, photodynamic tuning, and macro-optical interconnection can also be achieved. However, in practical applications, single-port optics do not fully meet the integration requirements of various optical systems, while OMCs do meet these requirements, which can be directly tapped into using two or more fibers. In addition, compared to OMs, OMCs not only have evanescent field transmission characteristics, but also have wavelength and polarization properties. The characteristics of this composite structure light field and functional devices that include the dual-waveguide coupling and evanescent field transmission, as well as the interaction of light fields with two-dimensional materials, have not yet been reported. This work will help to further understand the interaction characteristics of the light field and two-dimensional materials, and it is expected that new devices and application prospects can be obtained.

In view of this, a graphene-OMC integrated device (GOMC) was fabricated and its optical characteristics were investigated and characterized in this paper. Then, we used the GOMC in the development of adjustable multi-wavelength fiber lasers. By integrating the OMC with monolayer graphene, the polarization dependence of the OMC was enhanced. The developed GOMC had the characteristics of filtering, coupling, splitting, and polarization correlation. It is foreseeable that based on the GOMC, all-optical tunable mode lockers, filters, and mode-coupled selectors with better temperature control effects can be developed. In addition, it can provide optional functional devices and experimental platforms for the development of new fiber lasers and multi-dimensional light-field control research.

2. Fabrication and Characteristic Analysis of the GOMC

2.1. Fabrication of the GOMC

The GOMC proposed in this paper was made of a sandwich structure, in which graphene was first placed on a glass slide, and then the OMC was fixed on the glass slide. The manufacturing process of the GOMC is shown in Figure 1. First, to prevent the loss induced by the relatively high refractive index of the glass slide, a layer of MgF_2 with a low refractive index (1.376) and a thickness of $\approx$200 nm was evaporated onto it; then, a monolayer graphene sheet (after dissolving the Cu foil and removing the Polymethylmethacrylate (PMMA) film, 5×5 mm^2) was mechanically transferred to the surface of the MgF_2 [13]. The graphene film was directly synthesized using a chemical vapor deposition (CVD) method on the polycrystalline Cu substrate. Finally, the OMC was attached to the slide and the OMC waist area was aligned with the area where the graphene was located. Under the action of electrostatic forces, the waist region of the OMC was bonded to the graphene. The micrograph and structure of the junction of the OMC and graphene are shown in Figure 1d. The red scattering points at the GOMC waist region were caused by the contaminants during the sample transfer procedure (for the convenience of tracking display during the photographing process, red light was passed through the fiber). The OMC was fabricated by fusing and tapering two twisted conventional communication fibers, a method that was based on the improved flame-brushing method [6–8], which is comprised of a taper region and a uniform waist region. The schematic diagram of the composition OMC is shown in Figure 1c, where the P1, P2, P3, and P4 represent the four output ports of the OMC. Herein, the OMC sample for making the GOMC had a uniform waist length of 5 mm, a waist region diameter of 4 μm (i.e., the radius of the waist microfibers was 1 μm), and a lower insertion as low as 0.1 dB. This type of GOMC preparation process is simple and easy to operate, and the graphene film is sandwiched between the MgF$_2$ substrate and the OMC.

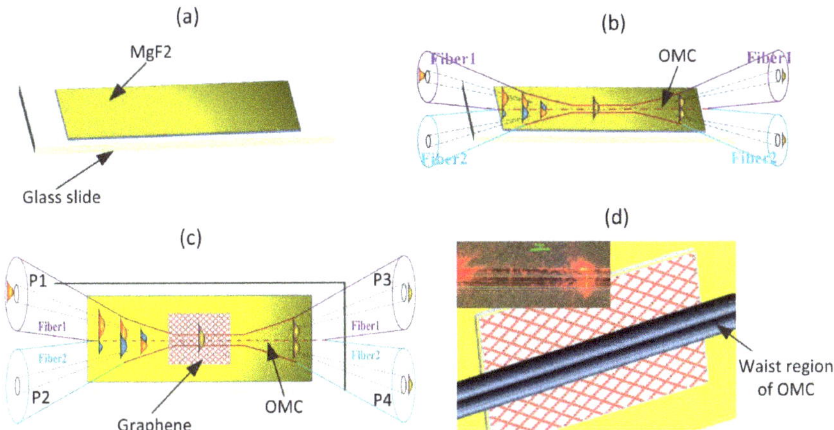

Figure 1. The preparation process of the GOMC: (**a**) evaporating MgF$_2$ onto the glass slide, (**b**) transfer graphene to the MgF$_2$ film, (**c**) the OMC was applied to a slide loaded with the graphene, and (**d**) the structure diagram and microscope photograph (insert image) of the OMC waist area and graphene bonding part.

2.2. Characteristic Analysis and Test of the GOMC

As can be seen from the GOMC fabrication process described above, the OMC sample used was based on the improved flame-brushing method. The total coupling of the OMC was composed of the transition region coupling and the uniform waist region coupling. According to the local coupled mode theory, the output light intensity of the two output ports of the OMC (P3 port, P4 port) can be expressed as [5–8]:

$$
\begin{cases}
P_3 = P_0 \cos^2\left(\int_0^l c(\lambda, n_2, n_3, z)dz \right) = P_0 \cos^2 \phi, \\
P_4 = P_0 \sin^2\left(\int_0^l c(\lambda, n_2, n_3, z)dz \right) = P_0 \sin^2 \phi,
\end{cases}
\tag{1}
$$

where P_0 is the input light power; l is the coupling length; $c(\lambda, n_2, z)$ is the coupling coefficient at wavelength λ and location z; and n_2 and n_3 are the refractive indexs (RIs) of fiber cladding (silica) and external environment, respectively. In fact, a thinner transition region and the uniform waist region of the OMC allow for its mode-coupling and spectral-transmission characteristics [5], and the GOMC proposed in this paper was made by bonding the waist region of the OMC with graphene. It can be seen from Equation 1 that the output spectrum of the two output ports (P3, P4) of the OMC should satisfy the conjugate symmetry relationship. The presence of the graphene absorbs part of the light-field energy, which is converted into the energy used to heat the optical fiber and the surrounding medium (MgF$_2$, glass, and air), resulting in the changes of n_2 and n_3. Therefore, the absorption of optical radiation in graphene could change (via a heating effect) the attenuation and coupling characteristics of the GOMC. It is worth noting that the presence of graphene increases the heat conduction area of the OMC waist region, transferring the heat to the surrounding media, such as air and glass, and thus helping to cool the OMC waist region [6].

The generation mechanism of the GOMC device's polarization properties was different from previous coupled-mode systems [17]. Its polarization correlation was not only related to the asymmetry of the OMC, but also to the light absorption of graphene, which gave the device more unique polarization characteristics. For the GOMC, the presence of graphene broke the cross-sectional symmetry of the combined device. Since the OMC had a large evanescent field similar to OM, part of the light field penetrates the graphene sheet into the surrounding environment (e.g., glass slide, MgF_2). The presence of graphene led to different absorption losses in different modes (light fields with different polarization directions), which mad the device exhibit a relative extinction ratio. This can be seen from the transmission spectrum test of the OMC sample and the GOMC sample, as shown in Figure 2. Therefore, compared with the OMC, the GOMC had different spectral transmission characteristics, which led to the output spectrum of the two output ports no longer satisfying the conjugate symmetry relationship. Therefore, the GOMC became a natural broad-spectrum polarization-dependent device.

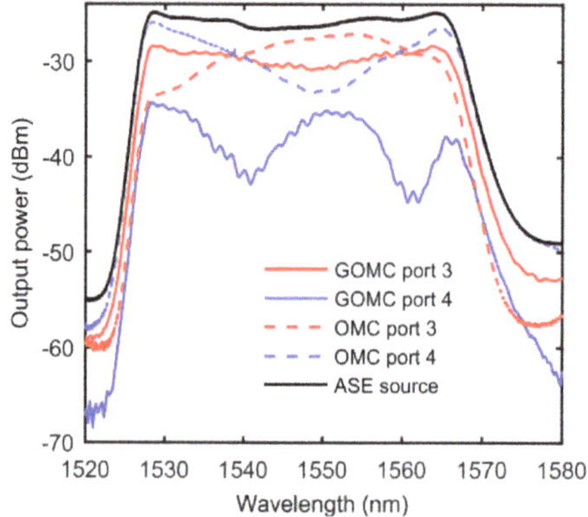

Figure 2. The transmission spectrum of the optical microfiber coupler (OMC) and the graphite-integrated OMC (GOMC) sample. ASE: Amplified Spontaneous Emission.

The polarization characteristics of fiber optic devices have a huge impact on their performance and system integration applications. In this work, the polarization and spectral transmission characteristics of the proposed GOMC composite device were experimentally studied, where the schematic diagram of the measurement system is shown in Figure 3. Because the amplified spontaneous emission (ASE) light source was non-polarized, we connected the source to the GOMC via an in-line fiber polarizer and a polarization controller (PC). Then, by adjusting the PC, different eigenmodes (polarization states) could be injected into the device to allow for the testing and analysis of the polarization transmission characteristics of the GOMC. The spectral transmission characteristics of the GOMC were analyzed using an optical spectrum analyzer (OSA; 600–1700 nm; Q8384, ADVANTEST, Japan; the resolution was 0.01 nm). In this study, the GOMC sample used in the experimental test had a splitting ratio of 3:1, corresponding to an output intensity ratio of 3 for the P3 and P4 ports. The device insertion loss of our GOMC was about 3 dB.

During the experiment, we connected the ASE directly to the GOMC to test the intrinsic spectral transmission characteristics of the device. The test results of the GOMC spectral transmission characteristics are shown in Figure 4, in which the red solid line and the dotted line are the characteristic spectra from the P3 and P4 port outputs, respectively. Comparing Figures 2 and 4, it can be seen that the characteristic spectra of the GOMC and the OMC were completely different, and neither met the conjugate symmetry nor regular periodic oscillation conditions. This shows that under the action of graphene and the glass slide, the mode transmission characteristics of the OMC waist region could not be considered two-mode interference. Moreover, during the OMC sample preparation, the twisting and tapering may have caused tolerances in the spacing of the two coupled microfibers, resulting in different coupling coefficients for the through port and the coupled port. However, from previous experience, the geometric asymmetry of the GOMC and the natural absorption of graphene were the main factors leading to the non-conjugation of the characteristic spectrum of the device.

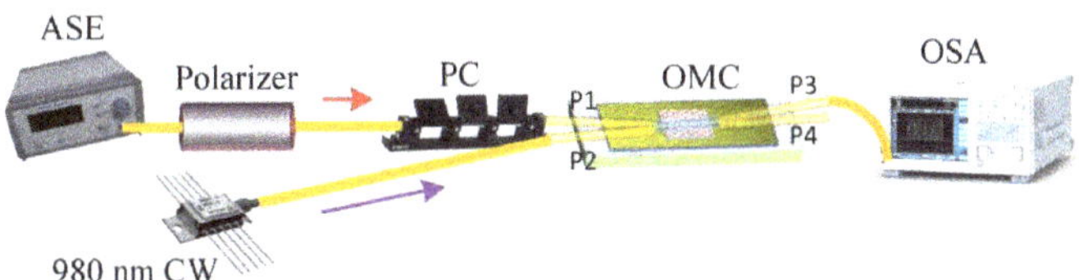

Figure 3. The schematic diagram of the GOMC transmission characteristic test system. CW: 980-nm Continuous Wave, OSA: Optical Spectrum Analyzer, PC: Polarization Controller.

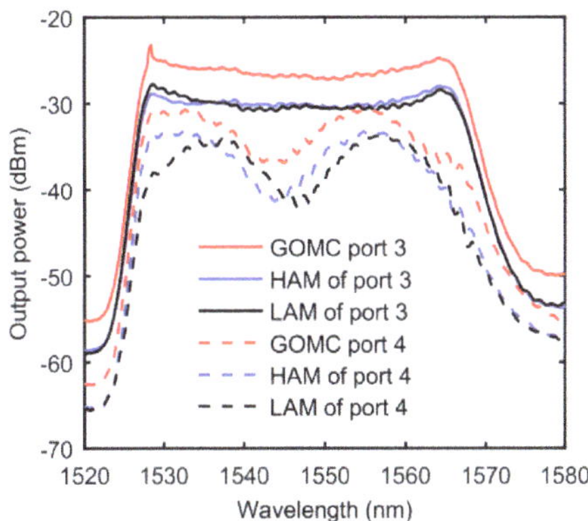

Figure 4. The spectral transmission characteristics of the GOMC. HAM: High-Absorption-Loss Mode, LAM: Low-Absorption-Loss Mode.

In addition, the ASE power input at the experiment was fixed at 10 mW. By adjusting the PC to inject two different polarization states (herein referred to as the low-absorption-loss mode (LAM) and the high-absorption-loss mode (HAM)) into the GOMC, the characteristic spectrum of the device was significantly shifted (i.e., red-shifted, the spectrum shift was due to the change of polarization state). When the GOMC was injected into the LAM and the HAM, the output intensity difference of the device was about 5 dB at 1550 nm, which was attributable to the broken symmetry of the GOMC geometry and the absorption of the graphene. It can be seen that the GOMC was very sensitive to the polarization and became a typical polarization-dependent device. In view of this, if the GOMC were to be combined with a PC, the tunable filtering function could be conveniently developed.

In order to further understand the temperature control characteristics of the device, the all-optical modulation method was employed to test and analyze the device. The internal light absorption heating effect was using to regulate the spectral transmission characteristics of the GOMC [6–8]. First, using the experimental system shown in Figure 3, different intensity "DC" 980-nm continuous pump light was injected into the device to test the "static" all-optical modulation characteristics of the polarization-dependent GOMC platform. During the experiment, the power of the ASE probe light injection device (10 mW) and the polarization control state of the PC were kept unchanged. The test results are shown in Figure 5. As the pump power was increased, the output spectra of the two output ports of the GOMC showed an overall attenuation, i.e., the device insertion loss rose while the modulated optical power was increased. The test results showed that when the pump power of 980 nm was increased from 2.43 mW to 34.9 mW, the overall attenuation of the output spectra of the P3 and P4 port were about 1.5 dB and 0.9 dB, respectively. However, the characteristic spectrum of the GOMC did not shift. This was different from the tunable filter introduced in Yang et al. [8]. Under the action of internal light absorption and heating, the GOMC could not achieve tunable filtering and its filtering characteristics were stable.

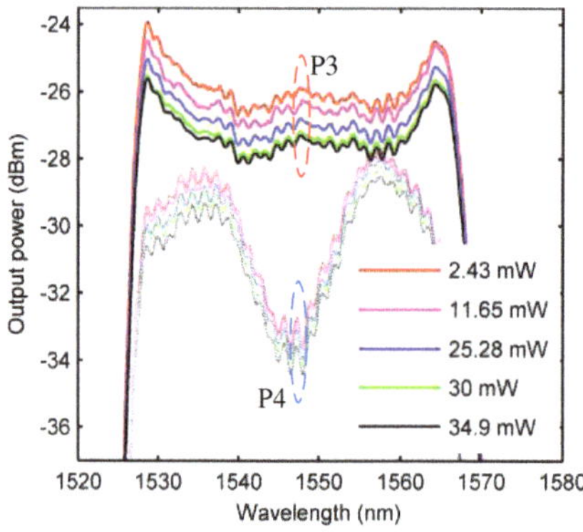

Figure 5. The all-optical modulation test results of the GOMC.

The light absorption of the GOMC was mainly from two aspects: waveguide intrinsic absorption of the fiber and graphene, and the PMMA of the graphene residue and adhesion of air pollutants during the sample preparation (the part of the energy of these absorbed light was converted into heat). Under the action of the thermo-optic effect, the OMC evanescent field ratio increased, which in turn caused the GOMC insertion loss to increase. However, the characteristic spectrum of the GOMC did not shift significantly during the modulation process. Therefore, it can be inferred that the extrinsic absorption of impurities was an important factor that caused the device loss to increase with the increase of the modulated optical power. In view of this, to improve the device stability and temperature control characteristics, various impurity contaminations introduced by the device fabrication process should be removed as much as possible. It should be noted that, unlike in Jin et al. [13], the detection optical power used in the experiment was relatively large (the intensity of the ASE injection device was equivalent to the modulated optical power); therefore, the intensity modulation phenomenon due to the saturated absorption characteristics of the graphene was not observed. According to the Pauli exclusion principle, it is necessary to use a short-wave, high-power pump light to control the long-wave weak probe light [13]. In fact, this experiment paid more attention to the effects of light absorption and heat on the transmission characteristics of the device to illustrate the thermal regulation characteristics of the GOMC.

The foregoing experiments showed that the characteristic spectrum of the GOMC did not shift significantly during the static pump light modulation. This came down to two factors. First, the two OMs of the OMC waist region had a large surface-to-volume ratio (the OMC's waist area can be regarded as two optical microfibers connected in parallel, as shown in Figure 1b), and its full adhesion to graphene further expanded the heat conduction surface [6]. Second, the excellent heat transfer characteristics of graphene could effectively release the heat generated by the light absorption conversion, thereby ensuring that the GOMC conformed to the waveguide boundary conditions. To further verify this inference, we also used the all-optical modulation method of Yang et al. [7], which used different frequencies of "dynamic" sinusoidal intensity signals to control the pump light injection into the GOMC to further test the light absorption and heat modulation characteristics of the device.

The experimental results showed that under the "dynamic" modulation of the pump light, the intensity modulation signal was not detected at the two output ports of the GOMC. This again verified the influence of graphene on the temperature characteristics of the device.

3. Fiber Laser Experiment and Discussion

The above analysis and test of the GOMC showed that the combined device had better temperature control characteristics and was very sensitive to the polarization, which could be used as a stable filter combined with the PC to realize tunable filtering. In view of this, the GOMC sample was used in the development of tunable multi-wavelength fiber lasers to further verify and develop the functions. The ring cavity fiber laser experimental test system is shown in Figure 6. The set-up used a 980-nm laser diode (LD) as the pump source, which passed through the 980/1550-nm wavelength division multiplexer (WDM) 980 nm port and common port, which was then connected to a 5-m erbium-doped fiber (EDF). Then, the EDF was connected to the P1 port of the GOMC via a 1550-nm polarization independent isolator (PI-ISO) that had a pump light absorption efficiency of approximately 13 dB/m. The isolator ensured that the laser was amplified, the output was clockwise within the laser cavity, and the laser stability was improved. The GOMC's P3 port was connected to the 1550-nm port of the 980/1550-nm WDM via a PC to form the complete fiber laser cavity loop. It can be seen from the above analysis and experiment that the adjustment of the PC could realize the regulation of the filtering characteristics of the GOMC, thereby realizing the multi-wavelength output of the fiber laser. The ring fiber laser had a total cavity length of about 7 m. The GOMC's P4 port (25% port) was used as the ring fiber laser output port, which could be connected to the OSA to analyze the characteristics of the laser output spectrum. The P2 port of the GOMC could be used to monitor the spontaneous emission and mode competition characteristics within the laser. During the experiment, the GOMC was packaged in a plastic box to minimize the impact of the environment on the laser system's stability.

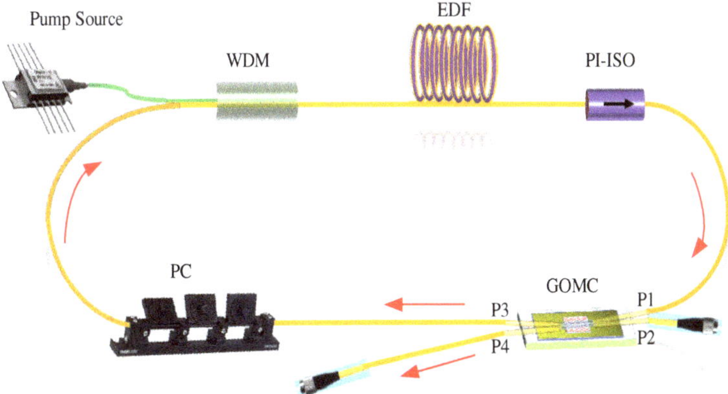

Figure 6. Experimental setup of the proposed laser based on the GOMC. EDF: Erbium-Doped Fiber, PI-ISO: Polarization Independent Isolator, WDM: Wavelength Division Multiplexer.

By adjusting the polarization state of the PC in the laser cavity, the laser could achieve single-wavelength, dual-wavelength, or even triple-wavelength outputs. In fact, more wavelengths of laser output state can be achieved through OMC structural optimization. For example, using a long or relatively thin waist sample, a dense frequency-comb-filtering function can be obtained to meet the needs of multi-wavelength laser development [8]. Figure 7 shows the output spectrum of the laser at single-wavelength and dual-wavelength outputs when the pump power was 200 mW. As can be seen in Figure 7, the signal-to-noise ratio (SNR) was about 40 ± 2 dB in the case of a single-wavelength output, and the illustration shows that the 3 dB line-width of the output laser was less than 30 pm. The test accuracy of the 3 dB line-width here was limited by the detection accuracy of the OSA.

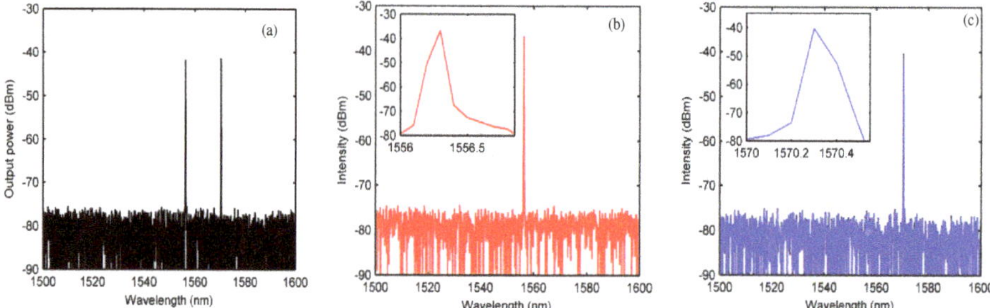

Figure 7. Test spectra of lasers in single-wavelength and dual-wavelength output states: (**a**) 1556-nm and 1570-nm output, (**b**) 1556-nm output, and (**c**) 1570-nm output.

In order to further understand the regulation of the polarization state of the PC using the laser wavelength output characteristics, we observed the laser output state after slightly selecting the PC's polarization control chip under the single-wavelength output state of the laser. The test results are shown in Figure 8. When the laser was operating in the short-wave (1556 nm) output state, as the PC's polarization control chip was rotated slightly clockwise, the output laser's center wavelength appeared red-shifted (as shown in Figure 8a). When the laser was operating in the long-wave (1570.3 nm) output state, as the PC's polarization control chip was rotated counterclockwise, the output laser's center wavelength appeared blue shift, and then gradually split into double peaks until the dual-wavelength output state was reached (Figure 8b).

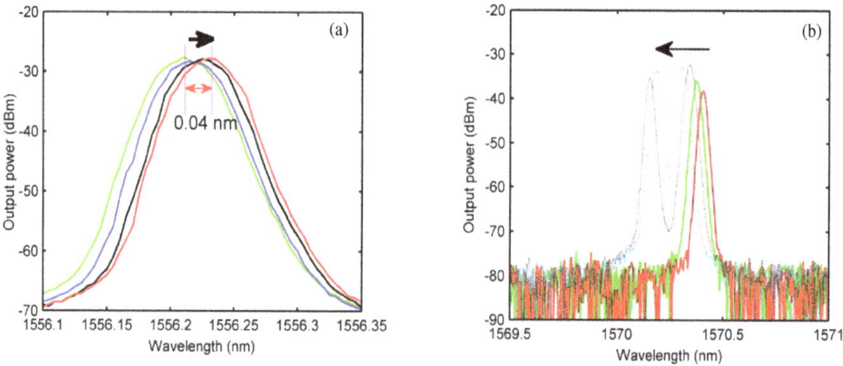

Figure 8. The laser output spectral test results during the fine-tuning of the PC: (**a**) rotation of the PC's polarization control sheet clockwise and (**b**) rotation of the control piece of the PC counterclockwise.

Comparing the polarization correlation test results of the aforementioned GOMC, it was not difficult to find that the polarization state change of the PC determined the overall filtering characteristics of the combined device with the GOMC, and thus the laser could be operated in different wavelength output modes. The polarization sensitivity and filtering function of the GOMC were also verified. In addition, the number of wavelengths output by this type of laser was mainly determined by GOMC filtering and insertion loss characteristics. By adjusting the length of the OMC waist region and the length of the region bonded to graphene, the GOMC samples with different filtering and insertion loss characteristics could be obtained. Therefore, the GOMC performed as a flexible device platform for the development of multi-wavelength fiber lasers.

During the experiment, the output spectral power stability of the laser was also tested. The test results are shown in Figure 9. In the single-wavelength and dual-wavelength modes of operation, the laser output power increased with the pump power, and the output spectrum was stable. At different pump optical powers, the laser's working center wavelength was stable without offset. The designed GOMC-based ring fiber laser could achieve stable, intensity-tunable, multi-wavelength-tunable laser output. This also indicated that the GOMC filtering characteristics were stable. The GOMC could accommodate the dual functions of the filter and the coupler in the laser cavity, which further simplified the laser system composition. Compared to other fiber laser systems, GOMC-based fiber lasers are simple, compact, low-cost, and easy to fabricate.

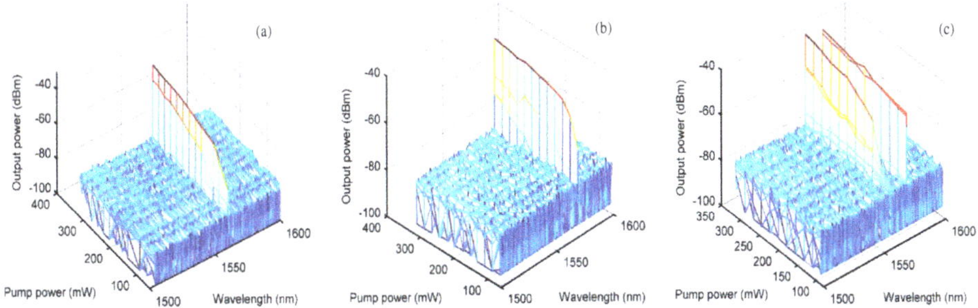

Figure 9. The output spectra of the laser under different 980-nm pump optical powers: (**a**) 1556-nm output, (**b**) 1570-nm output, and (**c**) 1556-nm and 1570-nm output.

Figure 10 shows the output efficiency of the constructed fiber laser. When the laser was using a single-wavelength output, the laser output power increased linearly with the pump light. When the laser was operating at a short wavelength (1556 nm), the laser threshold was approximately 65 mW and the output efficiency was approximately 2.7%. When the laser was operating at a long wavelength (1570 nm), the laser threshold was about 87 mW and the output efficiency was about 1.3% (as shown in Figure 10a). The threshold of the laser's dual-wavelength output was about 110 mW and the output power increased quadratically with the pump power. It can be seen that the short-wave output efficiency improved faster (as shown in Figure 10b). This indicated that the dual-wavelength output consumed a higher amount of pump light, but once the output was excited, the output efficiency increased rapidly with the pump power and was eventually as efficient as the single-wavelength output state. In addition, it can be seen from the test results that the laser cavity loss was still large. To meet practical applications, it is necessary to improve the output efficiency. The cavity loss of this laser was mainly determined by the insertion loss of the GOMC device (about 3 dB). In the sample preparation process, if the magazine residue on the graphene can be effectively removed and air pollution is avoided, a high-quality GOMC sample can be obtained. By optimizing the OMC parameter structure, the laser loss characteristics can be effectively improved.

We also tested the stability of the laser's dual-wavelength output state. During the experiment, the laboratory temperature (25 °C) and humidity (15%) were kept constant. When the pump power was 200 mW, the output state of the laser was monitored for 50 minutes. The test result is shown in Figure 11. The output power fluctuation of the 1556-nm center wave was less than 1 dBm, the output power fluctuation of the 1570-nm center wave was less than 0.8 dBm, and the two center wavelengths had no offset fluctuation. This power fluctuation was mainly due to the thermal noise of the laser system.

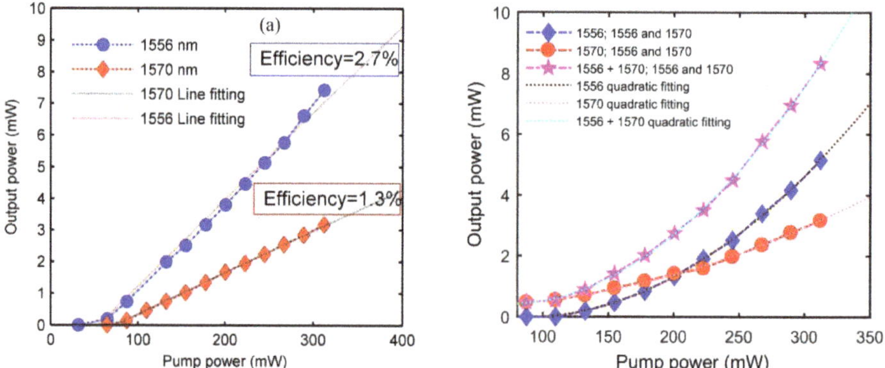

Figure 10. The output efficiency test results of the laser: (**a**) single wavelength output and (**b**) dual-wavelength output.

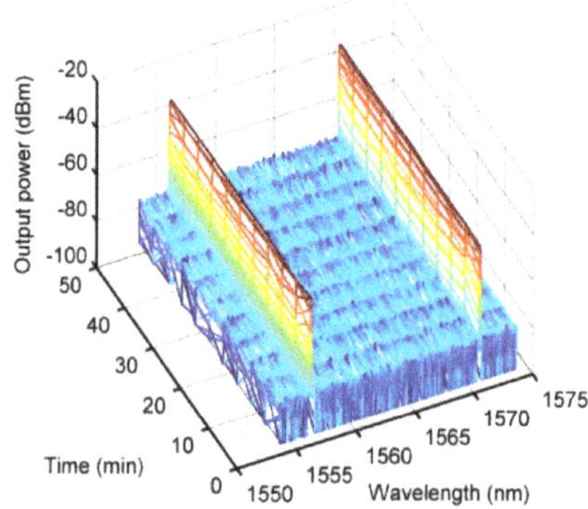

Figure 11. The stability test results of the laser.

The laser stability test showed that the proposed laser was better than the OMC-based fiber laser introduced in Harith [5]. This indicates that the presence of graphene can improve the thermal stability of the laser system to some extent. It also shows that the GOMC composite device has better thermal regulation characteristics and can be used as a key filter function device for laser development. In addition, this stable multi-wavelength laser can be widely used in communication, sensing, microwave photon radar, and other systems; therefore, it has great commercial value. It should be noted that when the GOMC was used as a light control function device, it needed to be packaged and protected. In an open environment, graphene itself will accumulate gases, impurities, and so on. Therefore, the GOMC can also be used as a sensing platform for gas and biosensor research.

4. Conclusions

We proposed and manufactured a GOMC combination device. Based on the characteristics analysis and testing of the device, it was applied to the development of multi-wavelength lasers. This device took full advantage of the high thermal conductivity of graphene and the advantages of OMC filtering. It had multi-port, evanescent field transmission; obvious polarization dependence; and excellent temperature adjustment characteristics. The experimental results demonstrated that the GOMC-based fiber laser could achieve single-wavelength and multi-wavelength regulated output. The 3-dB linewidth of the laser was less than 30 pm, the SNR was about 40 dB, and the output power fluctuation was less than 1 dB. The proven GOMC-based fiber laser featured the advantages of a compact structure, easy fabrication, and stable performance. In addition, the GOMC could perform as an alternative new experimental platform for light-field control, fiber laser development, and sensing applications.

Author Contributions: Y.Y. and H.C. led the device designing and testing. Z.Z., D.C., J.W., and Z.W. participated in the device fabrication and the experimental data processing. J.Y. and P.Y. wrote the manuscript. All authors have read and agreed to the published version of the manuscript.

Funding: This study was supported by the National Key R&D Program of China (2017YFB0405503), the National Natural Science Foundation of China (Nos. 61805278 and 61605249), the Equipment Pre-Research Field Foundation (61404140304), and the China Postdoctoral Science Foundation (2018M633704).

Acknowledgments: The authors would like to thank C.F. Zhang for help with the article writing.

Conflicts of Interest: The authors declare no conflict of interest.

References

1. Wei, F.; Mallik, A.K.; Liu, D.; Wu, Q.; Peng, G.-D.; Farrell, G.; Semenova, Y. Magnetic field sensor based on a combination of a microfiber coupler covered with magnetic fluid and a Sagnac loop. *Sci. Rep.* **2017**, *7*, 4725. [CrossRef] [PubMed]
2. Pu, S.; Luo, L.; Tang, J.; Mao, L.; Zeng, X. Ultrasensitive Refractive-Index Sensors Based on Tapered Fiber Coupler with Sagnac Loop. *IEEE Photon. Technol. Lett.* **2016**, *28*, 1073–1076. [CrossRef]
3. Sun, L.; Semenova, Y.; Wu, Q.; Liu, D.; Yuan, J.; Sang, X.; Yan, B.; Wang, K.; Yu, C.; Farrell, G. Investigation of Humidity and Temperature Response of a Silica Gel Coated Microfiber Coupler. *IEEE Photon. J.* **2016**, *8*, 1–7. [CrossRef]
4. Ding, M.; Wang, P.; Brambilla, G. A microfiber coupler tip thermometer. *Opt. Exp.* **2012**, *20*, 5402–5408. [CrossRef] [PubMed]
5. Ahmad, H.; Jasim, A.A. Fabrication and Characterization of 2 × 2 Microfiber Coupler for Generating Two Output Stable Multiwavelength Fiber Lasers. *IEEE J. L. Technol.* **2017**, *35*, 4227–4233. [CrossRef]
6. Yu, Y.Y.Y.; Bian, Q.B.Q.; Zhang, N.Z.N.; Lu, Y.L.Y.; Zhang, X.Z.X.; Yang, J.Y.J. Investigation on an all-optical intensity modulator based on an optical microfiber coupler. *Chin. Opt. Lett.* **2018**, *16*, 040605.
7. Yu, Y.; Bian, Q.; Guo, K.; Zhang, X.; Yang, J. Experimental Investigation on the Characteristics of All-Optical Modulation of Optical Microfiber Coupler. *IEEE Photon. J.* **2018**, *10*, 1–10. [CrossRef]
8. Yu, Y.; Bian, Q.; Wang, J.; Zhang, X.; Yang, J.; Liang, L. All-Optical Modulation Characteristics of a Microfiber Coupler Combined Sagnac Loop. *IEEE Photon. J.* **2018**, *11*, 1–11. [CrossRef]
9. Wan, H.; Wang, J.; Shen, Z.; Chen, Y.; Zhang, Z.; Wang, P.; Zhang, L. All fiber actively Q-switched Yb-doped laser with radially/azimuthally polarized beam generation. *L. Phys. Lett.* **2018**, *15*, 095102. [CrossRef]
10. Zhang, J.; Zhu, Z.; Liu, W.; Yuan, X.; Qin, S. Towards photodetection with high efficiency and tunable spectral selectivity: Graphene plasmonics for light trapping and absorption engineering. *Nanoscale* **2015**, *7*, 13530–13536. [CrossRef]
11. Koppens, F.H.L.; Mueller, T.; Avouris, P.; Ferrari, A.C.; Vitiello, M.S.; Polini, M. Photodetectors based on graphene, other two-dimensional materials and hybrid systems. *Nat. Nanotechnol.* **2014**, *9*, 780–793. [CrossRef]
12. Sun, Z.; Martinez, A.; Wang, F. Optical modulators with 2D layered materials. *Nat. Phot.* **2016**, *10*, 227–238. [CrossRef]
13. Chen, J.-H.; Zheng, B.-C.; Shao, G.-H.; Ge, S.-J.; Xu, F.; Lu, Y. An all-optical modulator based on a stereo graphene–microfiber structure. *Light. Sci. Appl.* **2015**, *4*, e360. [CrossRef]
14. Xin, W.; Liu, Z.-B.; Sheng, Q.-W.; Feng, M.; Huang, L.-G.; Wang, P.; Jiang, W.-S.; Xing, F.; Liu, Y.-G.; Tian, J.-G. Flexible graphene saturable absorber on two-layer structure for tunable mode-locked soliton fiber laser. *Opt. Express* **2014**, *22*, 10239–10247. [CrossRef] [PubMed]
15. Yin, J.; Li, J.; Chen, H.; Wang, J.; Yan, P.; Liu, M.; Liu, W.; Lu, W.; Xu, Z.; Zhang, W.; et al. Large-area highly crystalline WSe_2 atomic layers for ultrafast pulsed lasers. *Opt. Express* **2017**, *25*, 30020. [CrossRef] [PubMed]

16. Guo, B.; Wang, S.-H.; Wu, Z.-X.; Wang, Z.-X.; Wang, D.-H.; Huang, H.; Zhang, F.; Ge, Y.-Q.; Zhang, H. Sub-200 fs soliton mode-locked fiber laser based on bismuthene saturable absorber. *Opt. Express* **2018**, *26*, 22750–22760. [CrossRef] [PubMed]

17. Lobacha, I.A.; Babin, S.A.; Kablukov, S.; Podivilov, E.V.; Kurkov, A.S. Field distribution and mode interaction in twin-core fiber. *Laser Phys.* **2010**, *20*, 311–317. [CrossRef]

Letter

Metallic 2H-Tantalum Selenide Nanomaterials as Saturable Absorber for Dual-Wavelength Q-Switched Fiber Laser

Lingling Yang [1], Ruwei Zhao [1], Duanduan Wu [1], Tianxiang Xu [1,*], Xiaobiao Liu [2,*], Qiuhua Nie [1] and Shixun Dai [1]

[1] Laboratory of Infrared Materials and Devices, The Research Institute of Advanced Technologies, Ningbo University, Ningbo 315211, China; 1811082126@nbu.edu.cn (L.Y.); zhaoruwei@nbu.edu.cn (R.Z.); wuduanduan@nbu.edu.cn (D.W.); nieqiuhua@nbu.edu.cn (Q.N.); daishixun@nbu.edu.cn (S.D.)

[2] School of Sciences, Henan Agricultural University, Zhengzhou 450002, China

* Correspondence: xutianxiang@nbu.edu.cn (T.X.); liuxiaobiao@henau.edu.cn (X.L.); Tel.: +86-574-87609873 (T.X.)

Abstract: A novel 2H-phase transition metal dichalcogenide (TMD)–tantalum selenide ($TaSe_2$) with metallic bandgap structure is a potential photoelectric material. A band structure simulation of $TaSe_2$ via ab initio method indicated its metallic property. An effective multilayered $TaSe_2$ saturable absorber (SA) was fabricated using liquid-phase exfoliation and optically driven deposition. The prepared 2H–$TaSe_2$ SA was successfully used for a dual-wavelength Q-switched fiber laser with the minimum pulse width of 2.95 µs and the maximum peak power of 64 W. The repetition rate of the maximum pulse energy of 89.9 kHz was at the level of 188.9 nJ. The metallic 2H–$TaSe_2$ with satisfactory saturable absorbing capability is a promising candidate for pulsed laser applications.

Keywords: 2H–$TaSe_2$ nano-materials; metallic band structure; saturable absorber; dual-wavelength

Citation: Yang, L.; Zhao, R.; Wu, D.; Xu, T.; Liu, X.; Nie, Q.; Dai, S. Metallic 2H-Tantalum Selenide Nanomaterials as Saturable Absorber for Dual-Wavelength Q-Switched Fiber Laser. *Sensors* **2021**, *21*, 239. https://doi.org/10.3390/s21010239

Received: 11 December 2020
Accepted: 28 December 2020
Published: 1 January 2021

Publisher's Note: MDPI stays neutral with regard to jurisdictional claims in published maps and institutional affiliations.

1. Introduction

The increasing demand for pulsed lasers in the fields of scientific research and industrial processing has motivated researchers to explore a novel saturable absorber (SA). Materials can be divided into three categories depending on carrier concentration, namely, conductors, semiconductors, and insulators. At present, reported insulator SAs only include water [1] and alcohol [2]. The mechanism of saturable absorption is the electronic transition in vibrational energy levels of molecules. However, in-depth studies on insulator SAs are lacking. Related investigations on semiconductor SA have been conducted extensively [3–5]. Numerous SA materials, including SESAM, MXenes, black phosphorus, gold nanorods, and topological insulators, have been proposed to generate pulsed fiber laser. Semiconductor SAs are primarily constrained by their limited operating bandwidth because photon energies should be larger than the material bandgap, but conductor SAs can effectively solve this problem. At present, studies on conductor SAs commonly focus on metal nanoparticles [6,7], nanowires [8,9], and some special transition metal dichalcogenides (TMDs) [10].

TMDs have been regarded as graphene replacements due to their chemical structure of MX_2 (M and X denote the transition metal and chalcogen elements, respectively) and outstanding chemical, mechanical, and optoelectronic properties [11]. Given the abundance of M and X, TMDs have a large family with different characters. The chalcogen element X determines the stability and lattice parameter of TMDs, while the transition metal M influences electronic properties. Hence, TMDs can be categorized as semimetallic, metallic, or even superconducting materials [12,13]. Semiconductive TMDs, such as MoS_2, and WSe_2, have been extensively investigated [4]. However, metallic TMDs have rarely been explored. As a metallic TMD, 2H–tantalum selenide ($TaSe_2$) demonstrates unique magnetic, superconducting, and optical properties. Although the strain-induced ferromagnetism of monolayer TaSe$_2$, charge density wave, and functionality in logic circuits and switches have been investigated [14–16], reports on nonlinear optical properties of metallic 2H–TaSe$_2$ are limited.

The band structure and optical properties of 2H–TaSe$_2$ were investigated in this study. According to density functional theory, monolayer and multilayer 2H–TaSe$_2$ materials both exhibit a metallic electronic band structure. The fiber-integrated few-layered 2H–TaSe$_2$ SA was fabricated using liquid-phase exfoliation and optically driven deposition method. Open-aperture Z-scan measurement examined the saturable absorption property of 2H–TaSe$_2$. A dual-wavelength Q-switched fiber laser with a pulse energy and peak power of 188.9 nJ and 64 W, respectively, was demonstrated using the prepared 2H–TaSe$_2$ SA. The satisfactory nonlinear optical modulating ability of 2H–TaSe$_2$ proves its potential in novel optoelectronic applications.

2. Band Structure and Characterization of TaSe$_2$-SA

The atomic-layered structure of monolayer 2H–TaSe$_2$ is illustrated in Figure 1a. Similar to other TMDs, the transition metal (Ta) layer is sandwiched between two chalcogen (Se) layers. Our first-principle calculations were performed using Vienna Ab initio simulation package known as the VASP code [17–19]. The electronic–ion interaction is described using projector-augmented wave method [20]. Energy cutoff of plane waves was set to 520 eV. The electron-exchange correlation function was treated using generalized gradient approximation (GGA) in the form proposed by Perdew, Burke, and Ernzerhof (PBE) [21]. Both atomic positions and lattice vectors were fully optimized using conjugate gradient (CG) algorithm with an energy precision of 10–5 eV until the maximum atomic forces are smaller than 0.008 eV/Å. A vacuum region of approximately 15 Å was adapted to eliminate the interaction of two adjacent images. Brillouin zone (BZ) integration was sampled using a 17 × 17 × 1 k-mesh according to Monkhorst–Pack method [22]. The calculation above confirmed that 2H–TaSe$_2$ possesses a metallic and electronic band structure, as shown in Figure 1b. Varying band structures of 2H–TaSe$_2$ from 4 layers to 7 layers were further calculated using the same method, and the results are illustrated in Figure 2a–d. The corresponding atomic-layered structures are presented in Figure 2e. The calculated band structures demonstrated a small difference with the results of a previous report [23] because of the different settings in calculation parameters.

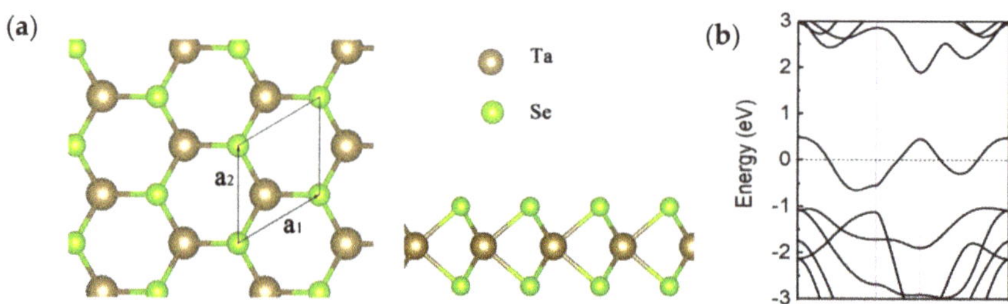

Figure 1. (a) Top (left) and side (right) view of monolayer TaSe$_2$ and (b) band structures of monolayer TaSe$_2$. Fermi level is set to 0 eV. G (0, 0, 0), M (0.5, 0, 0), and K (0.333333, 0.333333, 0) represent highly symmetric points in the reciprocal space.

The fiber-integrated few-layered TaSe$_2$ SA was fabricated using liquid-phase exfoliation and optically driven deposition method. Bulk TaSe$_2$ material was grinded into powder for intensive mixing with isopropyl alcohol (IPA). The initial mixture was then sonicated in an ultrasonic bath for approximately 8 h. Notably, this process has a suitable interval to prevent dispersion overheating. Finally, the as-prepared solution was centrifuged at 1500 rpm for 15 min to separate the large agglomeration, and the upper TaSe$_2$ supernatant was decanted and set aside.

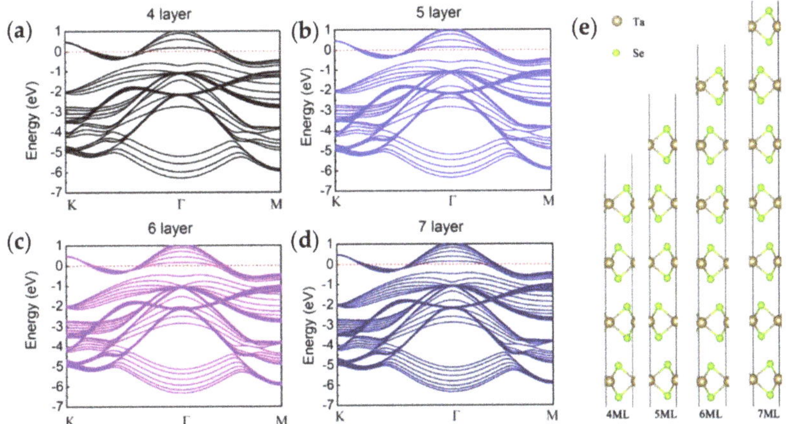

Figure 2. Band structures for multilayer TaSe$_2$: (**a**) four-, (**b**) five-, (**c**) six-, and (**d**) seven-layer TaSe$_2$; (**e**) corresponding atomic-layered structures of TaSe$_2$.

Homogeneously dispersed solution was dropped on a sapphire substrate to test the characterization of TaSe$_2$ nanosheets conveniently. Then, the substrate was dried under an infrared oven lamp for approximately 5 h to fabricate the TaSe$_2$ sample. Then the optically driven deposition method was applied to form a fiber-integrated few-layered TaSe$_2$ SA. Light with a wavelength and outpower of 980 nm and 60 mW, respectively, was directly irradiated onto the fiber. Fiber ferrule was dipped into TaSe$_2$ dispersion for 15 min. After drying for 24 h at room temperature, the fiber-integrated TaSe$_2$ SA was formed by connecting it with a clean one using an adaptor.

As-prepared TaSe$_2$ nanosheets were observed using AFM to investigate the morphology and layer character. The image area (40 μm × 40 μm) is illustrated in Figure 3a. A homodisperse solution was successfully fabricated. The corresponding cross-section analysis is shown in Figure 3b. The green dotted line denotes the average height, which showed that the thickness of the 2H–TaSe2 sample is approximately 3–5 nm; thus, the fabricated sample is a 4–7-layered structure (single-layered TaSe$_2$ shows a thickness of around 0.75 nm [24,25]).

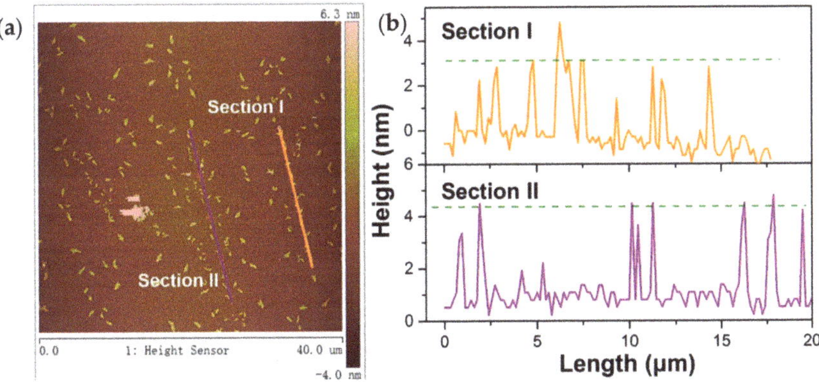

Figure 3. (**a**) Atomic force microscope (AFM) image of the 2H–TaSe$_2$ sample in a 40 × 40 μm region and (**b**) corresponding height.

The phase of the $TaSe_2$ sample was confirmed by investigating its Raman spectrum at room temperature using a laser source with an excitation of 532 nm (Renishaw inVia Raman microscope with the spectral resolution of 1 cm^{-1}). The 2H–$TaSe_2$ sample in the metallic phase demonstrates hexagonal syngony of D_{6h}^4 space group. A_{1g} and E_{1g} (E_{2g}) Raman active modes represent the out-of-plane and in-plain vibrational modes, respectively. Figure 4a shows that Raman peaks located at 140, 207, and 234 cm^{-1} correspond to E_{1g}, E_{2g}, and A_{1g} modes, respectively. Compared with previous reports, some negligible shifts of peak position and intensity caused by the difference in layer numbers of $TaSe_2$ nanosheets were observed [25,26]. Linear transmission spectra of the $TaSe_2$ sample and sapphire substrate in Figure 4b were examined using a spectrophotometer (U-3500) to show the broadband absorption at infrared band of metallic nanomaterial.

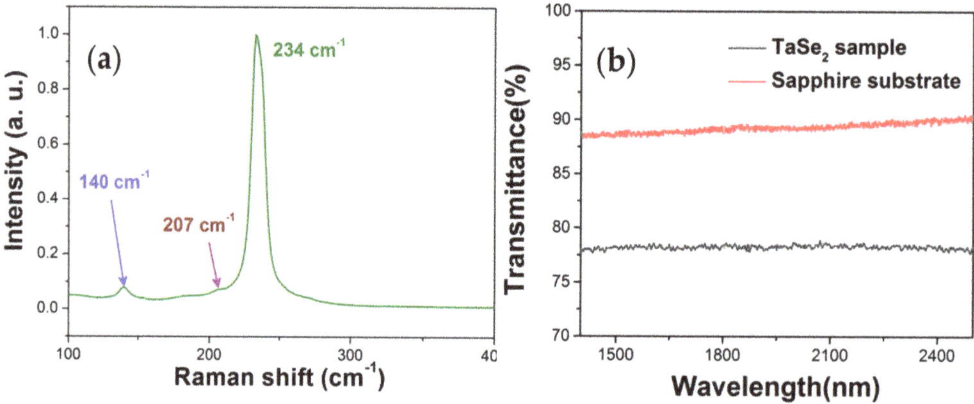

Figure 4. (**a**) Raman data of 2H-$TaSe_2$ nanosheets. (**b**) Linear transmittance of the sample and sapphire substrate.

An open-aperture Z-scan was applied to assess the nonlinear optical character of 2H–$TaSe_2$. A homemade pulsed fiber amplifier (wavelength, pulse width, and repetition rate of 1560 nm, 15 ps, and 1 MHz, respectively) was used to measure the Z-scan curve as follows (inset of Figure 5) [27]:

$$T(z) = \sum_{m=0}^{\infty} \frac{\left(-\alpha_{NL} I_0 L_{eff}\right)^m}{(m+1)^{1.5}\left[(z/z_0)^2+1\right]^m} \approx 1 - \frac{\alpha_{NL} I_0 L_{eff}}{2\sqrt{2}\left[(z/z_0)^2+1\right]} + \frac{\left(\alpha_{NL} I_0 L_{eff}\right)^2}{3\sqrt{3}\left[(z/z_0)^2+1\right]^2} \quad (1)$$

where L_{eff} is the effective length, α_{NL} is the nonlinear optical coefficient, I_0 is the power intensity, z_0 is the Rayleigh length, and T is the normalized transmittance. The nonlinear optical coefficient α_{NL} was (7.3 ± 0.7) cm/GW, according to the fitting result, which is comparable with the findings of other mature 2D materials [28]. According to the change in beam radius of the Gauss beam, the saturable absorption curve can be extracted based on Z-scan data, as illustrated in Figure 5. The modulation depth and saturation intensity were separately calculated at 9.6% and 12.7 µJ/cm^2, respectively, through data fitting [27]. The results showed that the metallic 2H–$TaSe_2$ sample can be a SA candidate for generating pulse lasers.

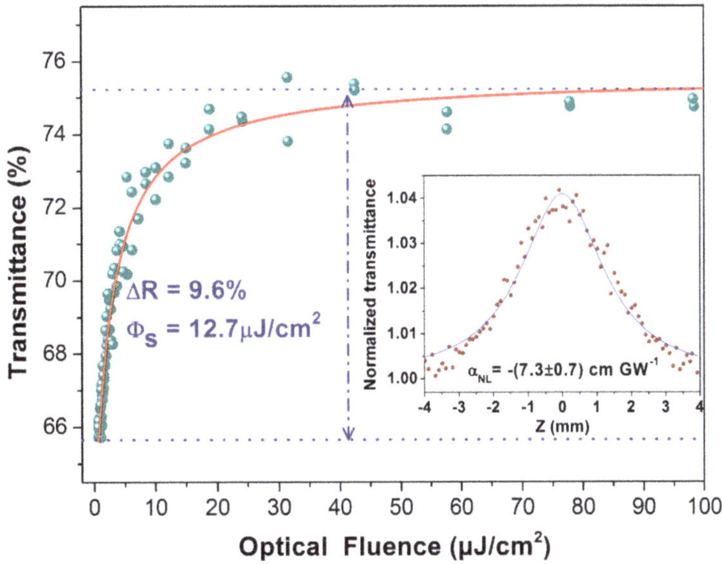

Figure 5. Saturable absorption curve of 2H-TaSe$_2$. (Note: The inset shows the open-aperture Z-scan data).

3. Experimental Setup

The Q-switched all-fiber laser cavity based on TaSe$_2$ SA is illustrated in Figure 6. The laser cavity length, including the 0.5 m-long and highly erbium-doped (EDF, LIEKKI: Er110-4-125) and else-tailed (single-mode fiber, SMF) fibers, was approximately 7.5 m. As a commercial 980 nm laser diode (LD) with a maximum power of 550 mW, the pump source is used via 980/1550 nm wavelength division multiplexing (WDM) to pump EDF with a dispersion parameter of -12 ps/nm/km. A polarization controller (PC) and a polarization-independent isolator (PI-ISO) were applied to optimize cavity birefringence and ensure the unidirectional operation of the laser cavity, respectively. Moreover, 20% of the energy of an optical coupler (OC) is used to output laser signal. Given that the standard SMF shows a dispersion parameter of 18 ps/nm/km, the net cavity dispersion is calculated at -0.153 ps^2.

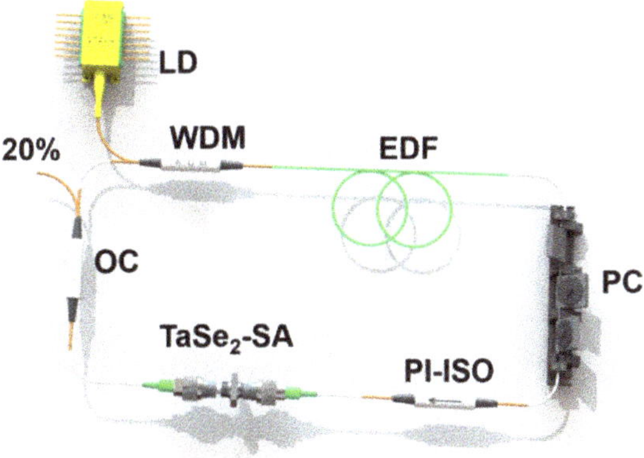

Figure 6. Experimental setup of the Q-switched fiber laser.

4. Results and Discussion

A clean ferrule without deposition was first inserted into the fiber cavity to prove the optical modulation capability of the TaSe$_2$ nanomaterial. As a result, Q-switch was absent, except continuous wave (CW), regardless of the pump power change or PC tuning. The output power was measured with a power meter (Thorlabs, PM100D), and its change versus pump power without TaSe$_2$ is illustrated in Figure 7a (green dots). The slope efficiency was 5.06% with the maximum output power of 23.8 mW under a pump power of 500 mW. The maximum output power decreased to 16.9 mW with a slope efficiency of 3.73% when the prepared TaSe$_2$ SA was inserted into the cavity due to the insertion loss induced by TaSe$_2$ SA. The results indicated the successful deposition of TaSe$_2$ materials onto the fiber ferrule.

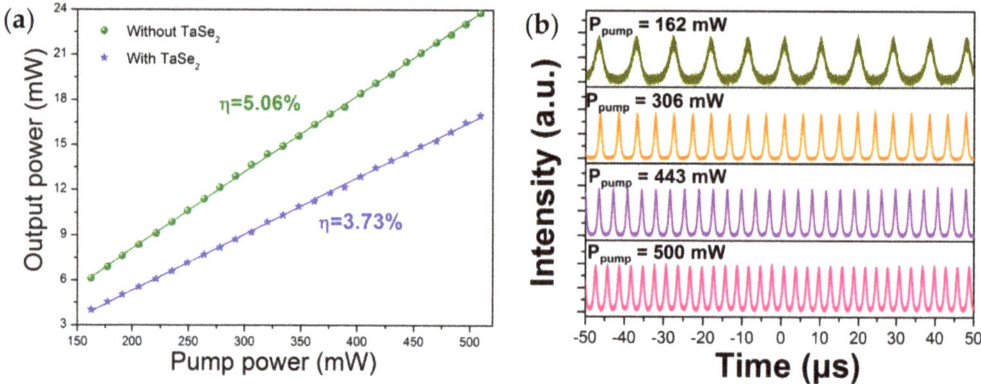

Figure 7. (**a**) Output power versus pump power with TaSe$_2$ (green dots) and without TaSe$_2$ (blue stars). (**b**) Stable pulse trains with different pump power values.

Stable and typical Q-switched pulses were observed when the pump power was increased to 160 mW. Figure 7b shows the output pulse trains with different pump power values. The evidently stable output pulses with the maximum pump power (500 mW) indicated the high thermal damage threshold of the metallic TaSe$_2$ material. When the pump power increased from 160 to 500 mW, the rate repetition increased from 28.4 to 89.9 kHz but the pulse width reduced from 9.69 to 2.95 μs. The minimum pulse with a Gauss-pulse profile assumed and pulse width (repetition rate) that depend on the pump power are depicted in Figure 8a,b, respectively. The change mechanism can be explained by the population inversion. The increase in electron accumulation on the upper energy level with increasing pump power shortens the rising and following time of big pulse formed by the sudden release of stored energy released suddenly. As a result, the pulse width reduced and the rate repetition increased with the increment of pump power. The pulse energy and peak power at different pump power values can be easily calculated, as shown in Figure 8c. The maximum pulse energy and peak power were 188.9 nJ and 64 W, respectively, at an average output power of 16.9 mW.

The output spectrum is illustrated in Figure 8d. A dual wavelength with spectral separation of 24.1 nm was detected. Methods of multiwavelength fiber laser generation, such as subzero treatment [29], Mach–Zehnder interferometry element [30–32], and highly nonlinear materials [33,34], have been extensively investigated. Two oscillation wavelength peaks were located at 1532.2 and 1556.3 nm while the oscillation wavelength of the CW state demonstrated only one peak at 1564.6 nm in this study. Therefore, the generation of dual-wavelength Q-switched lasers was caused by the high nonlinearity of the nanomaterial [33,34].

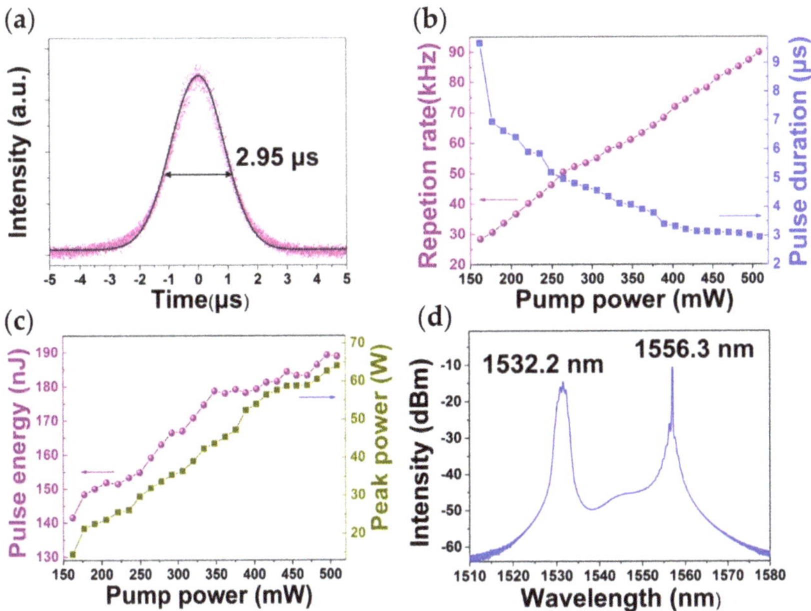

Figure 8. Laser performance: (**a**) Minimum pulse profile with a Gaussian-fitting standard deviation of 7.36×10^{-4}. (**b**) Pulse width and repetition rate versus pump power. (**c**) Pulse energy and peak power versus pump power. (**d**) Output spectrum with a pump power of 500 mW.

Finally, we explored a 2 h output power at a pump power of 500 mW (Figure 9a). The standard deviation of measured output power is 0.004 mW and the corresponding RMS (standard deviation/average output power) is 0.02%. Output spectra versus time at a pump power of 500 mW was also monitored (Figure 9b). The absence of evident vibrations indicated the excellent environmental stability of the Q-switched EDF lasers. Besides, Table 1 summarized the performance of Q-switched fiber laser using common semiconductor TMDs (including ternary TMD) materials as saturable absorber. By contrast, the 2H-TaSe$_2$ exhibited a comparable optical modulation ability. In view of the metallic band-structure, it could be deduced the 2H-TaSe$_2$ would be an except full-wave-band optical modulator.

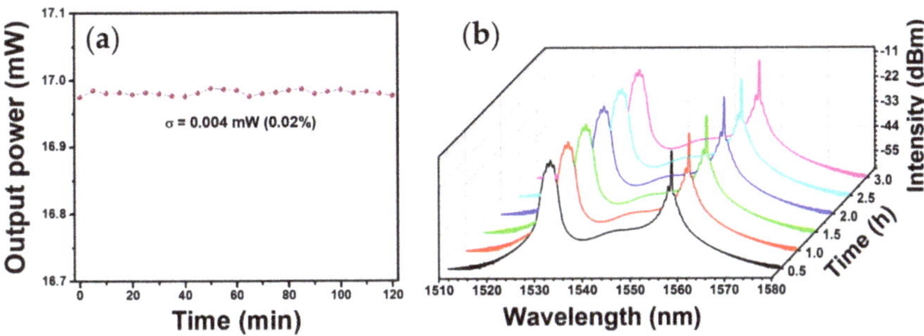

Figure 9. Long-term stability of (**a**) output power with the standard deviation of 0.004 mW and (**b**) output spectra at a pump power of 500 mW.

Table 1. Typical Q-switched EDF lasers using common transition metal dichalcogenides (TMDs) as saturable absorbers (SAs).

Materials	Modulation Depth	Saturation Intensity	Minimum Pulse Width (μs)	Pulse Energy (nJ)	Ref.
MoS_2	29%	$4.53\ MW/cm^2$	6	150	[35]
WS_2	7.7%	$342.6\ MW/cm^2$	0.1549	68.5	[36]
$MoSe_2$	6.73%	$132.5\ MW/cm^2$	4.04	365.9	[37]
WSe_2	7.17%	$7\ MW/cm^2$	1	29	[38]
$MoWSe_2$	19.7%	$18.9\ μJ/cm^2$	1.53	151.4	[27]
$2H\text{-}TaSe_2$	9.6%	$12.7\ μJ/cm^2$	2.95	188.9	This work

5. Conclusions

A novel $2H\text{–}TaSe_2$ with a metallic electronic band structure was used as a SA to generate Q-switched fiber laser. The nonlinear optical coefficient, modulation depth, and saturation intensity were $(7.3 \pm 0.7)\ cm/GW$, 9.6%, and $12.7\ μJ/cm^2$, respectively, via open-aperture Z-scan measurement. We demonstrated a dual-wavelength Q-switched fiber laser with a minimum pulse width of 2.95 μs based on the prepared $2H\text{–}TaSe_2$ SA. The maximum pulse energy of 188.9 nJ and maximum peak power of 64 W were calculated at the maximum repetition rate of 89.9 kHz. The excellent nonlinear optical modulating ability of $2H\text{–}TaSe_2$ verifies its potentiality in novel optoelectronic applications as a metallic TMD material.

Author Contributions: Investigation, L.Y.; writing—original draft preparation, R.Z. and T.X.; writing—review and editing, D.W. and X.L.; supervision, Q.N. and S.D. All authors have read and agreed to the published version of the manuscript.

Funding: This work is supported by National Natural Science Foundation of China (61905125, 61905124), China Postdoctoral Science Foundation (2018M642387), and K. C. Wong Magna Fund in Ningbo University.

Institutional Review Board Statement: Not applicable.

Informed Consent Statement: Not applicable.

Data Availability Statement: Data sharing is not applicable to this article.

Conflicts of Interest: The authors declare no conflict of interest.

References

1. Xian, T.; Zhan, L.; Gao, L.; Zhang, W.; Zhang, W. Passively Q-switched fiber lasers based on pure water as the saturable absorber. *Opt. Lett.* **2019**, *44*, 863–866. [CrossRef] [PubMed]
2. Wang, Z.Q.; Zhan, L.; Wu, J.; Zou, Z.; Zhang, L.; Qian, K.; He, L.; Fang, X. Self-starting ultrafast fiber lasers mode-locked with alcohol. *Opt. Lett.* **2015**, *40*, 3699–3702. [CrossRef] [PubMed]
3. Keller, U. Recent developments in compact ultrafast lasers. *Nat. Cell Biol.* **2003**, *424*, 831–838. [CrossRef] [PubMed]
4. Guo, B. 2D noncarbon materials-based nonlinear optical devices for ultrafast photonics. *Chin. Opt. Lett.* **2018**, *16*, 020004. [CrossRef]
5. Sun, Z.; Martinez, A.; Wang, F. Optical modulators with 2D layered materials. *Nat. Photon.* **2016**, *10*, 227–238. [CrossRef]
6. Wu, D.; Peng, J.; Cai, Z.; Weng, J.; Luo, Z.; Chen, N.; Xu, H. Gold nanoparticles as a saturable absorber for visible 635 nm Q-switched pulse generation. *Opt. Express* **2015**, *23*, 24071–24076. [CrossRef]
7. Pang, C.; Li, R.; Li, Z.; Dong, N.; Cheng, C.; Nie, W.; Böttger, R.; Zhou, S.; Wang, J.; Chen, F. Lithium Niobate Crystal with Embedded Au Nanoparticles: A New Saturable Absorber for Efficient Mode-Locking of Ultrafast Laser Pulses at 1 μm. *Adv. Opt. Mater.* **2018**, *6*, 1800357. [CrossRef]
8. Liu, W.-J.; Liu, M.L.; Lin, S.; Liu, J.C.; Lei, M.; Wu, H.; Dai, C.Q.; Wei, Z.-Y. Synthesis of high quality silver nanowires and their applications in ultrafast photonics. *Opt. Express* **2019**, *27*, 16440–16448. [CrossRef]
9. Liu, M.; Liu, W.-J.; Hou, H.; Ouyang, Y.; Lei, M.; Wei, Z.-Y. Silver nanowires with different concentration for Q-switched fiber lasers. *Opt. Mater. Express* **2020**, *10*, 187–197. [CrossRef]
10. XSun, X.; Shi, B.; Wang, H.; Lin, N.; Liu, S.; Yang, K.; Zhang, B.; He, J. Optical Properties of 2D 3R Phase Niobium Disulfide and Its Applications as a Saturable Absorber. *Adv. Opt. Mater.* **2020**, *8*, 1901181. [CrossRef]
11. Wang, Q.H.; Kalantar-Zadeh, K.; Kis, A.; Coleman, J.N.; Strano, M.S. Electronics and optoelectronics of two-dimensional transition metal dichalcogenides. *Nat. Nanotechnol.* **2012**, *7*, 699–712. [CrossRef] [PubMed]

12. Xi, X.; Wang, Z.; Zhao, W.; Park, J.-H.; Law, K.T.; Berger, H.; Forró, L.; Shan, J.; Mak, K.F. Ising pairing in superconducting NbSe$_2$ atomic layers. *Nat. Phys.* **2016**, *12*, 139–143. [CrossRef]

13. Soluyanov, A.A.; Gresch, D.; Wang, Z.; Wu, Q.; Troyer, M.; Dai, X.; Bernevig, B.A. Type-II Weyl semimetals. *Nature* **2015**, *527*, 495–498. [CrossRef] [PubMed]

14. Samnakay, R.; Wickramaratne, D.; Pope, T.R.; Lake, R.K.; Salguero, T.T.; Balandin, A.A. Zone-Folded Phonons and the Commensurate-Incommensurate Charge-Density-Wave Transition in 1T-TaSe$_2$ Thin Films. *Nano. Lett.* **2015**, *15*, 2965–2973. [CrossRef] [PubMed]

15. Manchanda, P.; Sharma, V.; Yu, H.; Sellmyer, D.J.; Skomski, R. Magnetism of Ta dichalcogenide monolayers tuned by strain and hydrogenation. *Appl. Phys. Lett.* **2015**, *107*, 032402. [CrossRef]

16. Renteria, J.; Samnakay, R.; Jiang, C.; Pope, T.R.; Goli, P.; Yan, Z.; Wickramaratne, D.; Salguero, T.T.; Khitun, A.G.; Lake, R.K.; et al. All-metallic electrically gated 2H-TaSe$_2$ thin-film switches and logic circuits. *J. Appl. Phys.* **2014**, *115*, 034305. [CrossRef]

17. Kresse, G.; Hafner, J. Ab initio molecular dynamics for liquid metals. *Phys. Rev. B* **1993**, *47*, 558–561. [CrossRef]

18. Kresse, G.; Hafner, J. Ab initio molecular dynamics for open-shell transition metals. *Phys. Rev. B* **1993**, *48*, 13115–13118. [CrossRef]

19. Kresse, G.; Furthmüller, J. Efficiency of ab-initio total energy calculations for metals and semiconductors using a plane-wave basis set. *Comput. Mater. Sci.* **1996**, *6*, 15–50. [CrossRef]

20. Blochl, P.E. Projector augmented-wave method. *Phys. Rev. B* **1994**, *50*, 17953–17979. [CrossRef]

21. Perdew, J.P.; Burke, K.; Ernzerhof, M. Generalized Gradient Approximation Made Simple. *Phys. Rev. Lett.* **1996**, *77*, 3865–3868. [CrossRef] [PubMed]

22. Monkhorst, H.J.; Pack, J.D. Special points for Brillouin-zone integrations. *Phys. Rev. B* **1976**, *13*, 5188–5192. [CrossRef]

23. Tsoutsou, D.; Aretouli, K.E.; Tsipas, P.; Marquez-Velasco, J.; Xenogiannopoulou, E.; Kelaidis, N.; Giamini, S.A.; Dimoulas, A. Epitaxial 2D MoSe$_2$ (HfSe$_2$) Semiconductor/2D TaSe$_2$ Metal van der Waals Heterostructures. *ACS Appl. Mater. Interfaces* **2016**, *8*, 1836–1841. [CrossRef] [PubMed]

24. Kuchinskii, E.Z.; Nekrasov, I.A.; Sadovskii, M.V. Electronic structure of two-dimensional hexagonal diselenides: Charge density waves and pseudogap behavior. *J. Exp. Theor. Phys.* **2012**, *114*, 671–680. [CrossRef]

25. Hajiyev, P.; Cong, C.; Qiu, C.; Yu, T. Contrast and Raman spectroscopy study of single- and few-layered charge density wave material: 2H-TaSe$_2$. *Sci. Rep.* **2013**, *3*, 2593. [CrossRef] [PubMed]

26. Wang, J.; Guo, C.; Guo, W.; Wang, L.; Shi, W.; Chen, X. Tunable 2H–TaSe$_2$ room-temperature terahertz photodetector. *Chin. Phys. B* **2019**, *28*, 046802. [CrossRef]

27. Yang, L.; Yan, B.; Zhao, R.; Wu, D.; Xu, T.; Nie, Q.; Dai, S. Output energy enhancement in a dual-wavelength Q-switched fiber laser based on a ternary MoWSe$_2$ saturable absorber. *Opt. Fiber Technol.* **2020**, *57*, 102214. [CrossRef]

28. Wang, K.; Feng, Y.; Chang, C.; Zhan, J.; Wang, C.; Zhao, Q.; Coleman, J.N.; Zhang, L.; Blau, W.J.; Wang, J. Broadband ultrafast nonlinear absorption and nonlinear refraction of layered molybdenum dichalcogenide semiconductors. *Nanoscale* **2014**, *6*, 10530–10535. [CrossRef]

29. Yamashita, S.; Hotate, K. Multiwavelength erbium-doped fibre laser using intracavity etalon and cooled by liquid nitrogen. *Electron. Lett.* **1996**, *32*, 1298–1299. [CrossRef]

30. Soltanian, M.R.K.; Amiri, I.; Alavi, S.E.; Ahmad, H. Dual-Wavelength Erbium-Doped Fiber Laser to Generate Terahertz Radiation Using Photonic Crystal Fiber. *J. Light. Technol.* **2015**, *33*, 5038–5046. [CrossRef]

31. Soltanian, M.R.K.; Ahmad, H.; Khodaie, A.; Amiri, I.; Ismail, M.F.I.M.F.; Harun, S.W. A Stable Dual-wavelength Thulium-doped Fiber Laser at 1.9 μm Using Photonic Crystal Fiber. *Sci. Rep.* **2015**, *5*, 14537. [CrossRef] [PubMed]

32. Ahmad, H.; Soltanian, M.R.K.; Pua, C.H.; Zulkifli, M.Z.; Harun, S.W. Narrow Spacing Dual-Wavelength Fiber Laser Based on Polarization Dependent Loss Control. *IEEE Photon. J.* **2013**, *5*, 1502706. [CrossRef]

33. Zhao, R.; He, J.; Su, X.; Nie, H.; Yang, K.; Wang, Y.; Sun, X.; Zhang, B. Tunable High-Power Q-Switched Fiber Laser Based on BP-PVA Saturable Absorber. *IEEE J. Sel. Top. Quantum Electron.* **2017**, *24*, 1–5. [CrossRef]

34. Zhao, X.; Zheng, Z.; Liu, L.; Liu, Y.; Jiang, Y.; Yang, X.; Zhu, J. Switchable, dual-wavelength passively mode-locked ultrafast fiber laser based on a single-wall carbon nanotube modelocker and intracavity loss tuning. *Opt. Express* **2011**, *19*, 1168–1173. [CrossRef]

35. Chen, J.-H.; Deng, G.-Q.; Yan, S.-C.; Li, C.; Xi, K.; Xu, F.; Lu, Y.-Q. Microfiber-coupler-assisted control of wavelength tuning for Q-switched fiber laser with few-layer molybdenum disulfide nanoplates. *Opt. Lett.* **2015**, *40*, 3576–3579. [CrossRef]

36. Chen, H.; Chen, Y.; Yin, J.; Zhang, X.; Guo, T.; Yan, P. High-damage-resistant tungsten disulfide saturable absorber mirror for passively Q-switched fiber laser. *Opt. Express* **2016**, *24*, 16287–16296. [CrossRef]

37. Chen, B.; Zhang, X.; Wu, K.; Wang, H.; Wang, J.; Chen, J. Q-switched fiber laser based on transition metal dichalcogenides MoS$_2$, MoSe$_2$, WS$_2$, and WSe$_2$. *Opt. Express* **2015**, *23*, 26723–26737. [CrossRef]

38. Guo, C.; Chen, B.; Wang, H.; Wu, K.; Chen, J. Investigation on Stability of WSe$_2$-PVA in an All Polarization Maintaining Q-Switched Fiber Laser. *Chin. J. Laser.* **2017**, *44*, 0703018.

sensors

Communication

1.1-μm Band Extended Wide-Bandwidth Wavelength-Swept Laser Based on Polygonal Scanning Wavelength Filter

Gi Hyen Lee [1,†], Soyeon Ahn [1,†], Jinhwa Gene [2] and Min Yong Jeon [1,2,*]

1   Department of Physics, College of Natural Sciences, Chungnam National University, 99 Daehak-ro Yuseong-gu, Daejeon 34134, Korea; gtit5de@naver.com (G.H.L.); ahnsoyen5@naver.com (S.A.)
2   Institute of Quantum Systems (IQS), Chungnam National University, 99 Daehak-ro Yuseong-gu, Daejeon 34134, Korea; genejh@gmail.com
*   Correspondence: myjeon@cnu.ac.kr
†   These authors contributed equally to this work.

Abstract: We demonstrated a 1.1-μm band extended wideband wavelength-swept laser (WSL) that combined two semiconductor optical amplifiers (SOAs) based on a polygonal scanning wavelength filter. The center wavelengths of the two SOAs were 1020 nm and 1140 nm, respectively. Two SOAs were connected in parallel in the form of a Mach-Zehnder interferometer. At a scanning speed of 1.8 kHz, the 10-dB bandwidth of the spectral output and the average power were approximately 228 nm and 16.88 mW, respectively. Owing to the nonlinear effect of the SOA, a decrease was observed in the bandwidth according to the scanning speed. Moreover, the intensity of the WSL decreased because the oscillation time was smaller than the buildup time. In addition, a cholesteric liquid crystal (CLC) cell was fabricated as an application of WSL, and the dynamic change of the first-order reflection of the CLC cell in the 1-μm band was observed using the WSL. The pitch jumps of the reflection band occurred according to the electric field applied to the CLC cell, and instantaneous changes were observed.

Keywords: wavelength-swept laser; fiber lasers; semiconductor optical amplifier; dynamic measurement; dynamic optical fiber sensors

Citation: Lee, G.-H.; Ahn, S.; Gene, J.; Jeon, M.-Y. 1.1-μm Band Extended Wide-Bandwidth Wavelength-Swept Laser Based on Polygonal Scanning Wavelength Filter. *Sensors* **2021**, *21*, 3053. https://doi.org/10.3390/s21093053

Academic Editor: Vittorio M. N. Passaro

Received: 29 March 2021
Accepted: 26 April 2021
Published: 27 April 2021

Publisher's Note: MDPI stays neutral with regard to jurisdictional claims in published maps and institutional affiliations.

1. Introduction

A wavelength-swept laser (WSL) is a light source that can continuously change its narrow linewidth wavelength over a wide wavelength range at high speed [1–19]. Owing to its wide wavelength band and fast wavelength scanning speed, it is primarily used as a light source for optical coherence tomography (OCT) systems in biophotonics [2–12]. In addition, the output of the WSL has a one-to-one correspondence in the spectral and temporal domains; thus, it has been widely applied as a light source for dynamic fiber optic sensors that measure dynamic changes in wavelength [13–18]. WSLs can be implemented using a variety of methods. Among them, the polygonal scanning wavelength filter-based WSL [2–4,17–20] and the Fabry-Perot tunable filter (FFP-TF)-based WSL [5–13,21–23] have been studied most actively. In addition, an electro-optical tunable filter (EOTF) [24] and acousto-optic tunable filter (AOTF) [25,26] have been researched. FFP-TF has the advantage of optical alignment, which can be easily implemented because all elements comprise a pigtailed optical fiber; however, the center wavelength of the filter is unstable, owing to thermal instability. Because a sinusoidal function is applied to the filter, the signal is nonlinear and requires complex signal processing [10,11]. An EOTF is not limited by a mechanical drive and fast reaction speed, and it operates linearly [24]. Therefore, the scanning bandwidth is small and the linewidth is relatively wider than that of other methods. An AOTF also has the advantage of no mechanical drive; however, the bandwidth of the maximum gain is not sufficiently wide and the scanning speed is slow [25,26]. Although the polygonal scanning wavelength filter is relatively bulky and is limited by mechanical driving, it is possible to easily change the scanning speed and scanning wavelength range by adjusting the rotation speed, diffraction grating angle, and the magnification of the telescope [2–4,17–20].

Most WSLs have been implemented for biophotonics imaging applications in the 1300-nm band [2–5,8–12]. In addition, many studies have utilized various wavelength bands, such as 850- [24,27–29], 1000- [30–32], 1500- [6,13,17,21,22,25,33–35], and 1700–2000-nm [36–38] bands, as light sources for OCT imaging, including optical fiber sensors. Research on WSLs typically focuses on obtaining a fast scanning speed [15,39–44] and scanning in a wide wavelength band [8,9,12,37] to improve imaging quality or the sensing dynamic range. MEMS-based WSL [45,46], dispersion tuned WSL [15,47–49], very short cavity WSL using FFP-TF [9], and Fourier domain mode locked WSL [5,6,11–13,16,22] were implemented to achieve a fast scanning speed. Methods to implement a wide scanning wavelength band include connecting two gain media in parallel and using a semiconductor optical amplifier (SOA) with a wide gain area [4,8,12]. If a wide scanning wavelength band in a WSL is implemented in the 1-μm band, high resolution can be realized as an OCT light source or the dynamic measurement range can be increased in the optical fiber sensor system.

In this study, we successfully demonstrated, for the first time to our knowledge, a >228-nm wideband WSL around 1.1-μm band based on a polygonal scanning wavelength filter using two SOAs. Two SOAs were combined in parallel as a Mach-Zehnder interferometer in the laser cavity. This enabled a wider wavelength scanning band by combining the adjacent wavelength bands of the two SOAs. In addition, the characteristics of the scanning bandwidth and average power were investigated with respect to the scanning speed of the WSL. As an application of the dynamic measurement of WSL, the phenomenon of the pitch jump was observed according to the intensity of the electric field applied to a cholesteric liquid crystal (CLC) cell and the observation of the instantaneous movement of the first-order reflection band on the oscilloscope was reported.

2. Experiments

The WSL is a wavelength-tunable laser that continuously and rapidly varies with time in a wide wavelength bandwidth. In the wavelength-tunable filter inserted into the laser cavity, only the maximum gain corresponding to the filter condition is fed back to the resonator when an amplified spontaneous emission (ASE) beam with a wide bandwidth is incident on the filter. By continuously changing these conditions, the WSL continuously oscillates over a wide bandwidth.

Figure 1 shows a schematic diagram of the experimental setup, in which two SOAs were connected in parallel in the form of a Mach-Zehnder interferometer to construct a single polygonal scanning wavelength filter-based WSL. This obtained a wider wavelength scanning band by combining the adjacent wavelength bands of the two SOAs [4,8,12]. When two SOAs are connected in series, a wide scanning band cannot be obtained because the gain of one SOA is absorbed by the other [4]. The broadband WSL consisted of two SOAs, two polarization controllers in front and behind each SOA, two 50:50 fiber couplers, an optical circulator, and a polygonal scanning wavelength filter; the last is indicated by the dotted box in Figure 1. The polygonal scanning wavelength filter contained a brazed diffraction grating, a telescope with two lenses, and a 36-facet polygonal scanning mirror. The telescope was comprised of two achromatic doublet lenses with a grating at the front focal plane of the first lens and a polygonal scanning axis on the back focal plane of the second lens. The parallel beam from the collimator was incident on a brazing diffraction grating and underwent diffraction of the first order, which was incident on the telescope and aligned along the optical axis. The diffracted wavelength components had different angles of convergence on the polygonal scanning mirror facet. Therefore, the polygonal scanning mirror only reflected the spectral components within a narrow resolution band that were vertically incident. The reflected wavelength component was fed back into the laser cavity. Because the polygonal scanning mirror rotated at a high speed, the lasing wavelength continuously varied within the gain band. If the polygonal scanning mirror was rotated in the direction of increasing wavelength, the energy was transmitted in a long wavelength, owing to the nonlinear effect of the SOA. Therefore, a higher output and narrower line width was obtained, compared with the opposite case. A 600-lines/mm diffraction grating was used, the angle of incidence was 47°, the angle of reflection was 3.5°, and the focal lengths of the two lenses were each 5 cm. The output from the WSL was monitored on an oscilloscope using a photodetector and an optical spectrum analyzer (OSA).

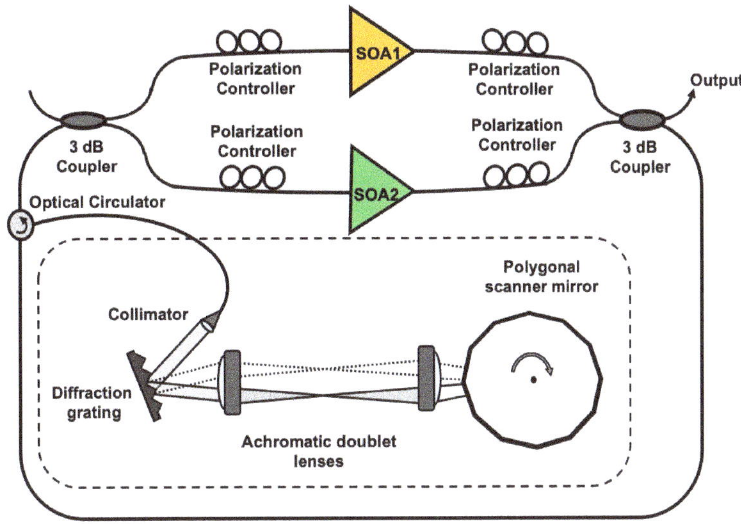

Figure 1. Schematic diagram of an experimental setup, in which two SOAs were connected in parallel in the form of a Mach-Zehnder interferometer.

In the experiment, a 1064 ± 100-nm broadband optical fiber coupler and a broadband circulator operating at 950–1100 nm were used to minimize the optical loss across the wide wavelength band. The center wavelengths of the ASE of SOA 1 and SOA 2 were 1020 nm and 1140 nm, respectively, and the 10-dB bandwidths were 114 nm and 57 nm, respectively, as shown in Figure 2.

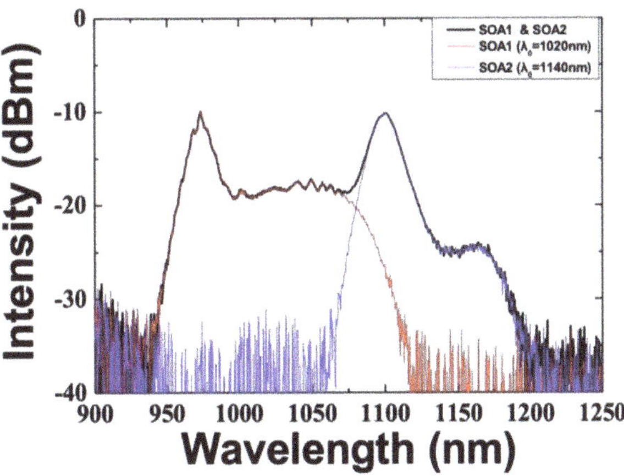

Figure 2. ASE spectra of two SOAs.

Figure 3a shows the optical spectrum of the WSL when only SOA 1 was connected in the Mach-Zehnder interferometer. The 10-dB bandwidth and average output power were 132 nm (from 959 nm to 1091 nm) and 11.46 mW, respectively. Figure 3b shows the optical spectrum of the WSL when only SOA 2 was connected in the Mach-Zehnder interferometer. The 10-dB bandwidth and average output power were 108 nm (from 1079 nm to 1187 nm) and 4.73 mW, respectively. Figure 3c shows the optical spectra output from the fabricated WSL by combining the two SOAs in parallel at a scanning speed of 1.8 kHz. The red and blue lines denote the optical spectrum when only SOA 1 or SOA 2 were connected to the Mach-Zehnder interferometer of the laser cavity, respectively. The optical spectrum of the WSL, obtained by connecting the two SOAs together, is represented as a black line. The 10-dB bandwidth and the average output power of the WSL with two SOAs were ~228 nm (from 959 nm to 1187 nm) and ~16.88 mW, respectively. The wavelength scanning resolution measured by OSA was 2 nm, but the same scanning bandwidth of 228 nm or more was obtained with 0.2-nm resolution. This is a significantly wider scanning bandwidth compared with that of the ASE with two SOAs, and the output spectrum of the WSL achieved a relatively uniform amplitude. This can be achieved by controlling the pump current of each SOA and controlling the polarization appropriately using the polarization controllers in the laser cavity. The instantaneous linewidth was determined to be 0.11 nm in this laser cavity. Figure 3d shows the output of the WSL measured using the oscilloscope. This corresponds to the optical spectrum of the WSL, shown in Figure 3c. In the oscilloscope, the time interval for a scanning bandwidth pulse was measured by 440 μs, and a period was measured as 560 μs. The period corresponds to free spectral range (FSR) of the wavelength filter. By inversely converting the wavelength scanning range of 228 nm corresponding to the time interval of 440 μs of the scanning bandwidth pulse, the wavelength range corresponding to 560 μs was inversely estimated. The obtained FSR was approximately 290 nm. The measurement error was ~2.5% due to the resolution of the oscilloscope. The FSR of the wavelength filter can be calculated using the following equation [2,50]:

$$(\Delta\lambda)_{\text{FSR}} = p \, (\cos\beta_0)\frac{F_2}{F_1}\theta_0 \, , \tag{1}$$

where p is the grating pitch, $\theta_0 = 2\pi/36$ is the facet-to-facet polar angle of the polygonal scanner mirror, β_0 is the angle between the optical axis of the telescope and the grating normal, and F_1 and F_2 are the focal lengths of the two lenses in the telescope. Using Equation (1), the calculated FSR of the wavelength filter is 290 nm. The FSR measured on the oscilloscope was approximately 290 nm, which is similar to the theoretical value. The measured 10-dB wavelength scanning range of the WSL was ~228 nm; therefore, the duty cycle was approximately 78.6%. In order to obtain the desired FSR and to operate normally, it is necessary to appropriately adjust various variables shown in Equation (1). However, in order to obtain sufficient wavelength scanning, the FSR must be greater than the wavelength scanning band of the light source.

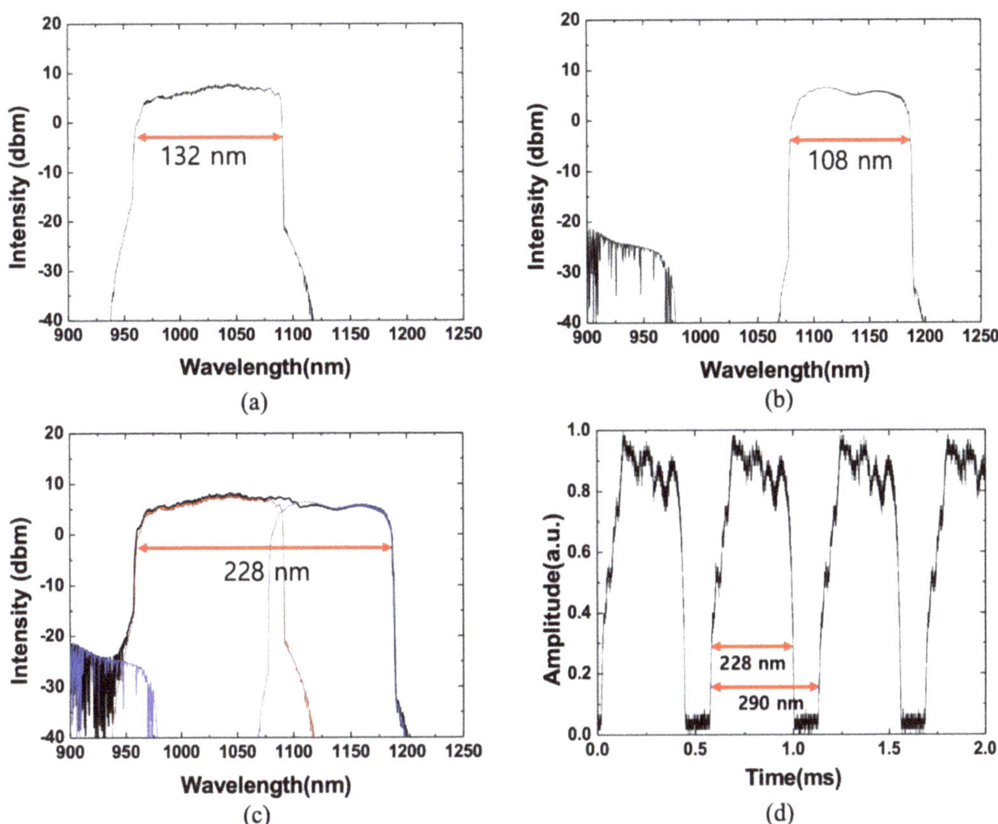

Figure 3. (**a**) Optical spectrum of the WSL when only SOA 1 was connected in the Mach-Zehnder interferometer; (**b**) optical spectrum of the WSL when only SOA 2 was connected in the Mach-Zehnder interferometer; (**c**) optical spectra output and (**d**) corresponding temporal output from the WSL by combining two SOAs in parallel in the Mach-Zehnder interferometer.

Figure 4a,b shows the optical spectra of WSL with a linear scale when only SOA 1 or SOA 2 were independently connected, and Figure 4c shows the optical spectrum of the WSL with a linear scale when SOA1 and SOA2 were combined in a Mach-Zehnder interferometer configuration. If the output spectra of the two SOAs overlapped, the interference between the two laser outputs can cause intensity noise. However, in the case of a parallel configuration, beating noise may occur if the resonator lengths have exactly matched each other, but it can be eliminated by introducing an offset of the length between the two arms within the Mach-Zehnder interferometer [4,12]. Additionally, Figure 4d–f shows the output pulses in the temporal domain corresponding to Figure 4a–c. They show a one-to-one correspondence between the shape of the pulse signal in the temporal domain and the wavelength band in the spectral domain. Therefore, the dynamic optical properties of a material can be inferred from a wavelength signal by measuring the output pulse in the temporal domain using the WSL.

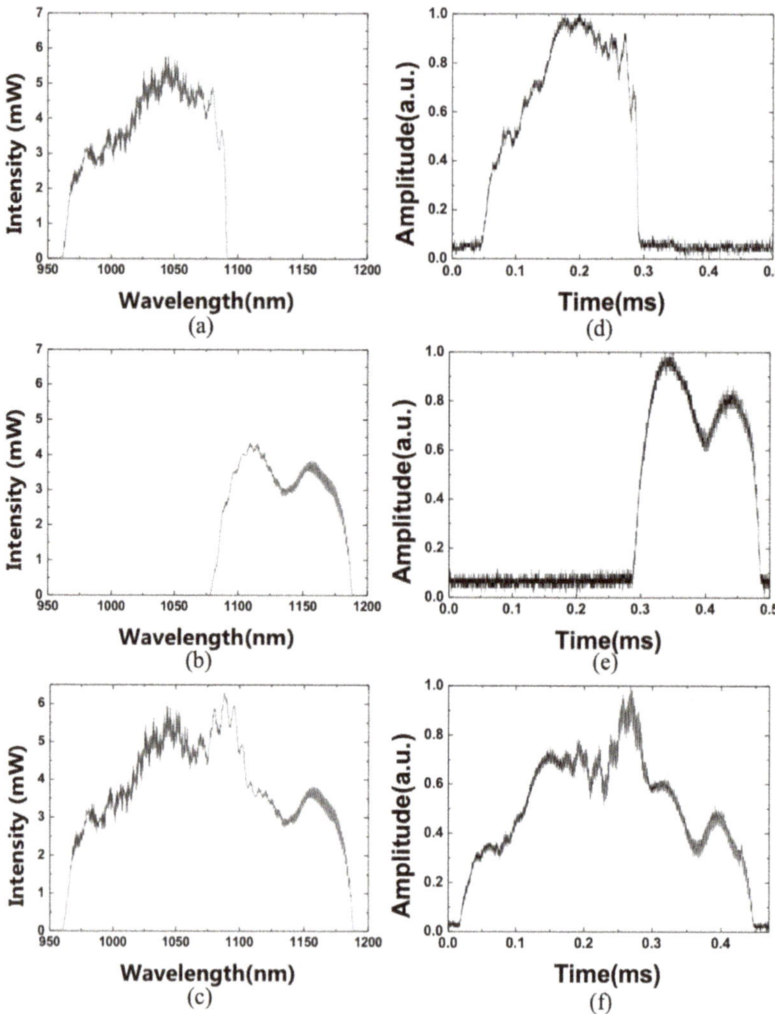

Figure 4. Optical spectrum of WSL with a linear scale (**a**) when only SOA 1 was connected, (**b**) when only SOA 2 was connected, and (**c**) when SOA1 and SOA2 were combined in a Mach-Zehnder interferometer configuration. (**d–f**) Output pulses in the temporal domain corresponding to (**a–c**).

To investigate the characteristics of the WSL, the change in the scanning bandwidth was measured while increasing the scanning speed. Figure 5a shows the variation in the spectra according to the scanning speed of the WSL. Up to 2 kHz, a 10-dB scanning bandwidth of the WSL achieved over ~228 nm. As the scanning speed increased, the scanning bandwidth gradually decreased. At approximately 8 kHz, the scanning bandwidth was reduced to ~220 nm, as shown in Figure 5b. Similarly, as shown in Figure 5c, as the scanning speed increased, the average optical output power gradually decreased. This occurred because the gain was not sufficiently obtained at an oscillation time smaller than the buildup time of the WSL.

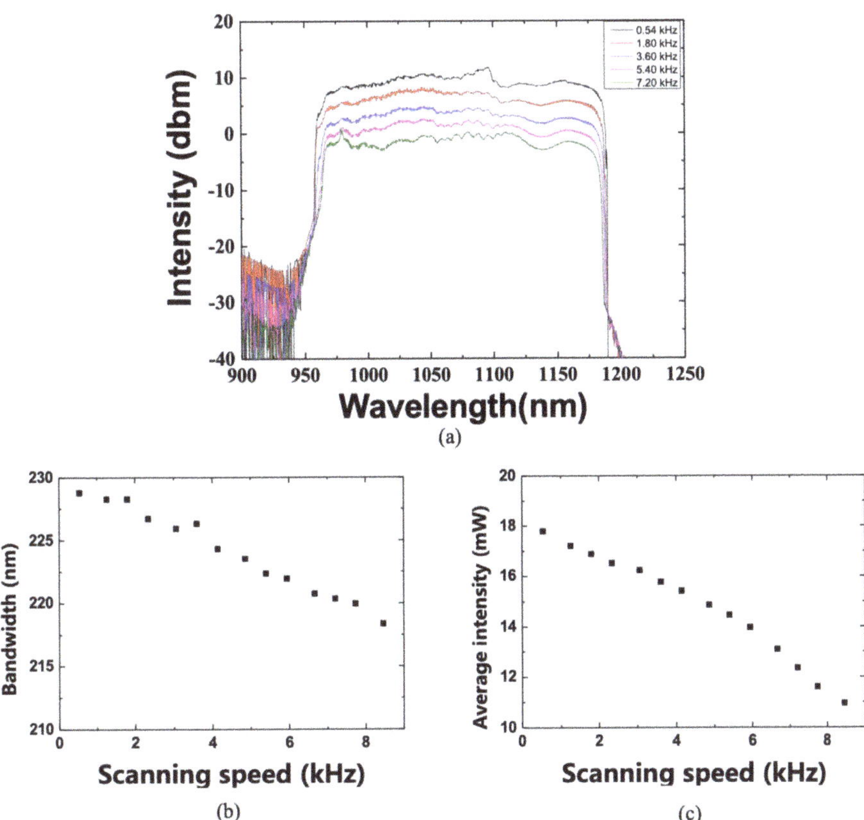

Figure 5. (a) Optical spectra, (b) 10-dB bandwidth, and (c) average optical power with respect to the WSL scanning speed.

As mentioned above, the output of WSL has a one-to-one correspondence in the spectral and temporal domain; therefore, it has been widely applied as a light source to measure dynamic changes in wavelength [13–18]. As a simple application, the dynamic variation of the first-order reflection spectrum from a CLC cell is measured by applying an electric field to the CLC cell in the 1-μm band. In the experiments, a nematic liquid crystal E7 and chiral dopant (R811) were mixed to produce a right-handed CLC. The chiral dopant concentration of the CLC cell was 13.92 wt%, and the calculated pitch was 678 nm. Figure 6 shows a photograph of the output signal of the WSL on an oscilloscope. The scanning speed was 1.8 kHz and the duty cycle was ~78.6%. The time intervals of 440 μs and 560 μs on the oscilloscope correspond to a 228-nm scanning bandwidth and a 290-nm FSR in the spectral domain, respectively.

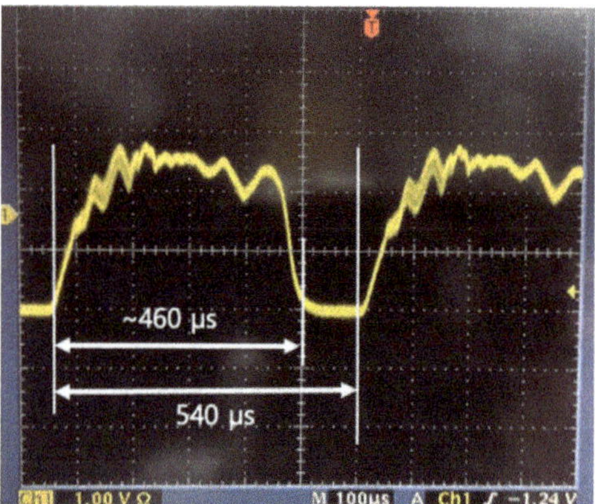

Figure 6. Photograph of the output signal of the WSL on an oscilloscope.

The beam from the WSL was incident on the CLC cell, and the transmission spectra for the first-order reflection with respect to the intensity of the electric field applied to the CLC cell is observed. Figure 7a–c shows the optical spectra of the reflection band when the applied electric field is 2.70, 3.42, and 3.77 $V_{rms}/\mu m$, respectively. The wavelength of the arrow indicated in the figures is the wavelength of the short-band edge of the reflection band. When an electric field of 2.70 $V_{rms}/\mu m$ or more was applied to the CLC cell, the short edge of the reflection band shifted to 1044 nm from 1022 nm. When the electric field applied to the CLC cell increased, the reflection band moved to longer wavelength discontinuously, owing to the pitch jump, which occurred instantaneously [19]. When an electric field of 3.42 $V_{rms}/\mu m$ or more was applied, the short edge shifted to 1071 nm, as shown in Figure 7b. Because the OSA's response to the wavelength change was slow, it was not easy to observe the abrupt change in the reflection band spectrum, owing to the instantaneous pitch jump. However, when the dynamic variations of the WSL were observed in the temporal domain, using an oscilloscope and high-speed photodetector, the process of changing the wavelength of the reflection band owing to the instantaneous pitch jump could be observed. Figure 7d–f shows the oscilloscope displays of the first-order reflection band when the applied electric field was 2.70, 3.42, and 3.77 $V_{rms}/\mu m$, respectively. These correspond to Figure 7a–c, respectively. On the oscilloscope, if the electric field applied to the CLC cell was continuously increased, it could be observed in real time that the short-band edge of the reflection band was instantaneously moved by the pitch jump at any moment. As an example, when the electric field increased from 3.42 to 3.77 $V_{rms}/\mu m$, the pitch jump occurred in the CLC cell. Supplementary Material Videos S1 and S2 show videos of the in situ variation of the first-order reflection band spectrum, owing to the dynamic pitch jump of the CLC cell, on the oscilloscope. Supplementary Material Video S1 shows a video in which the reflection band changes when the electric field applied to the CLC cell increases from 3.42 to 3.77 $V_{rms}/\mu m$ and Supplementary Material Video S2 shows the electric field applied to the CLC cell is increased from 3.77 to 4.13 $V_{rms}/\mu m$.

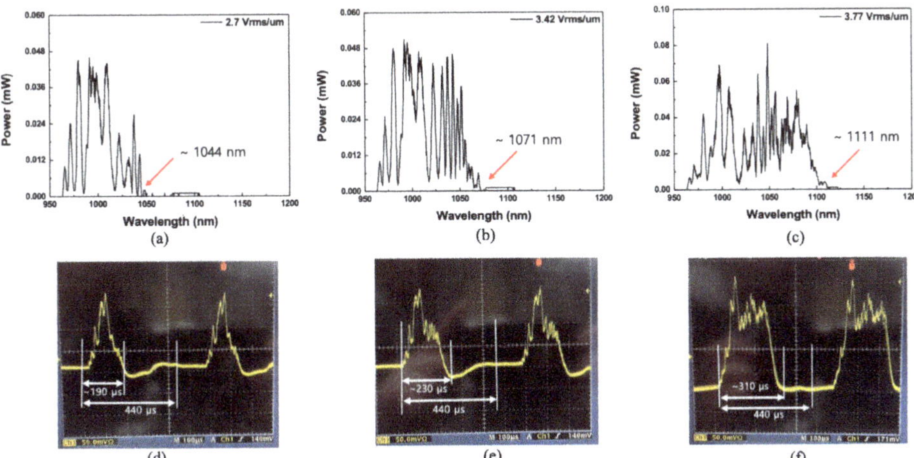

Figure 7. Optical spectra of the first-order reflection band from the CLC cell when the electric field was (**a**) 2.70, (**b**) 3.42, and (**c**) 3.77 $V_{rms}/\mu m$, and (**d**–**f**) the oscilloscope displays of the first-order reflection band corresponding to the applied electric fields in (**a**–**c**), respectively [Supplementary Videos S1 and S2].

Figure 8 shows the wavelength shift of the short-band edge on the OSA and oscilloscope when a pitch jump occurred with respect to the electric field applied to the CLC cell. In the figure, the error bar represents the error owing to the thickness of the line when measured using the oscilloscope. It can be observed that the wavelength shifts obtained by converting the time interval measured on the oscilloscope into a wavelength are notably consistent within the measurement error compared with that measured by the OSA. Therefore, if a WSL is used to measure the dynamic wavelength change, it can also easily determine the wavelength change in the OSA by measuring the waveform change using the oscilloscope and converting it into a wavelength.

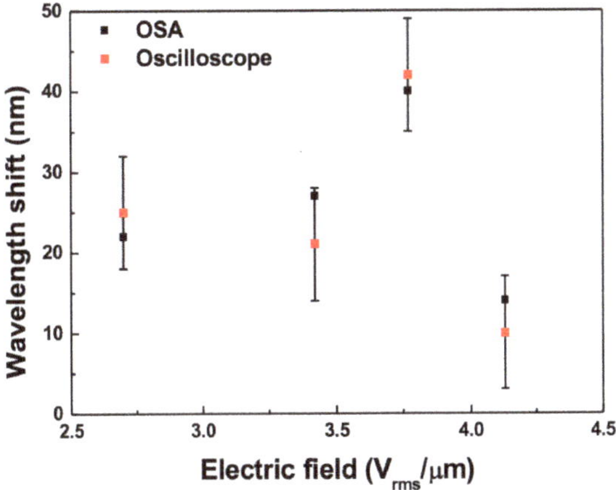

Figure 8. Wavelength shift of the short-band edge on the OSA and oscilloscope when a pitch jump occurs according to the electric field applied to the CLC cell.

3. Conclusions

We successfully demonstrated a wide-bandwidth WSL based on a polygonal scanning wavelength filter using two SOAs. By combining two SOAs in parallel in the form of a

Mach-Zehnder interferometer, we achieved a 10-dB bandwidth of ~228 nm (from 959 to 1187 nm). We also investigated the changes in the scanning bandwidth and average output power with respect to the scanning speed of the WSL. The bandwidth and average power of the WSL decreased because the oscillation time was smaller than the buildup time of the SOA. A CLC cell was fabricated to investigate the first-order reflection band, owing to the dynamic pitch jump, with respect to the electric field applied to the CLC cell in the 1100-nm band region. The instantaneous change in the reflection band of the CLC cell was due to the instantaneous pitch jump of the CLC. Moreover, the dynamic change in the reflection band of the CLC cell was confirmed by converting the instantaneous change of the waveform on the oscilloscope into the corresponding wavelength change. The wide scanning wavelength band in the 1.1-μm band of WSL is expected to be used as a high-resolution OCT light source or to increase the dynamic measurement range in a fiber optic sensor system.

Supplementary Materials: The following are available online at https://www.mdpi.com/article/10.3390/s21093053/s1. Video S1: A video in which the reflection band changes when the electric field applied to the CLC cell increased from 3.42 V_{rms}/μm to 3.77 V_{rms}/μm. Video S2: A video in which the reflection band changes when the electric field applied to the CLC cell increased from 3.77 V_{rms}/μm to 4.13 V_{rms}/μm.

Author Contributions: G.H.L. and S.A. conceived and designed the experiments; J.G. reviewed and analyzed the data; G.H.L. and S.A. contributed to writing—original draft preparation; M.Y.J. contributed to writing—review and editing; M.Y.J. supervised. All authors have read and agreed to the published version of the manuscript.

Funding: This research was supported by Chungnam National University of Korea and was supported by Basic Science Research Program through the National Research Foundation of Korea (NRF) funded by the Ministry of Science, ICT and future Planning (NRF-2019R1A2C1084933, NRF-2020R1A6A1A03047771) and was supported by a Korea Institute for Advancement of Technology (KIAT) grant funded by the Korea Government (MOTIE) (P0008458, The Competency Development Program for Industry Specialist).

Institutional Review Board Statement: Not applicable.

Informed Consent Statement: Not applicable

Data Availability Statement: Data available on request from the authors.

Conflicts of Interest: The authors declare no conflict of interest.

References

1. Yun, S.H.; Richardson, D.J.; Kim, B.Y. Interrogation of fiber grating sensor arrays with a wavelength-swept fiber laser. *Opt. Lett.* **1998**, *23*, 843–845. [CrossRef] [PubMed]
2. Yun, S.H.; Boudoux, C.; Tearney, G.J.; Bouma, B.E. High-speed wavelength-swept semiconductor laser with a polygon-scanner-based wavelength filter. *Opt. Lett.* **2003**, *28*, 1981–1983. [CrossRef] [PubMed]
3. Oh, W.Y.; Yun, S.H.; Tearney, G.J.; Bouma, B.E. 115 kHz tuning repetition rate ultrahigh-speed wavelength-swept semiconductor laser. *Opt. Lett.* **2005**, *30*, 3159–3161. [CrossRef] [PubMed]
4. Oh, W.Y.; Yun, S.H.; Tearney, G.J.; Bouma, B.E. Wide tuning range wavelength-swept laser with two semiconductor optical amplifiers. *IEEE Photonics Technol. Lett.* **2005**, *17*, 678–680. [CrossRef]
5. Huber, R.; Wojtkowski, M.; Fujimoto, J.G. Fourier domain mode locking (FDML): A new laser operating regime and applications for optical coherence tomography. *Opt. Express* **2006**, *14*, 3225–3237. [CrossRef]
6. Eigenwillig, C.M.; Wieser, W.; Todor, S.; Biedermann, B.R.; Klein, T.; Jirauschek, C.; Huber, R. Picosecond pulses from wavelength-swept continuous-wave Fourier domain mode-locked lasers. *Nat. Commun.* **2013**, *4*, 1848. [CrossRef]
7. Jirauschek, C.; Huber, R. Wavelength shifting of intra-cavity photons: Adiabatic wavelength tuning in rapidly wavelength-swept lasers. *Biomed. Opt. Express* **2015**, *6*, 2448–2465. [CrossRef]
8. Kassani, S.H.; Villiger, M.; Uribe-Patarroyo, N.; Jun, C.; Khazaeinezhad, R.; Lippok, N.; Bouma, B.E. Extended bandwidth wavelength swept laser source for high resolution optical frequency domain imaging. *Opt. Express* **2017**, *25*, 8255–8266. [CrossRef]
9. Jun, C.; Villiger, M.; Oh, W.-Y.; Bouma, B.E. All-fiber wavelength swept ring laser based on Fabry-Perot filter for optical frequency domain imaging. *Opt. Express* **2014**, *22*, 25805–25814. [CrossRef]

10. Huber, R.; Wojtkowski, M.; Taira, K.; Fujimoto, J.G. Amplified, frequency swept lasers for frequency domain reflectometry and OCT imaging: Design and scaling principles. *Opt. Exp.* **2005**, *13*, 3513–3528. [CrossRef]

11. Eigenwilling, C.M.; Biedermann, B.R.; Palte, G.; Huber, R. K-space linear Fourier domain mode locked laser and applications for optical coherence tomography. *Opt. Exp.* **2008**, *16*, 8916–8937. [CrossRef]

12. Jeon, M.Y.; Zhang, J.; Wang, Q.; Chen, Z. High-speed and wide bandwidth Fourier domain mode-locked wavelength swept laser with multiple SOAs. *Opt. Express* **2008**, *16*, 2547–2554. [CrossRef]

13. Jung, E.J.; Kim, C.-S.; Jeong, M.Y.; Kim, M.K.; Jeon, M.Y.; Jung, W.; Chen, Z. Characterization of FBG sensor interrogation based on a FDML wavelength swept laser. *Opt. Express* **2008**, *16*, 16552–16560. [CrossRef]

14. Isago, R.; Nakamura, K. A high reading rate fiber Bragg grating sensor system using a high-speed swept light source based on fiber vibrations. *Meas. Sci. Technol.* **2009**, *20*, 034021. [CrossRef]

15. Nakazaki, Y.; Yamashita, S. Fast and wide tuning range wavelength-swept fiber laser based on dispersion tuning and its application to dynamic FBG sensing. *Opt. Express* **2009**, *17*, 8310–8318. [CrossRef]

16. Lee, B.C.; Jung, E.-J.; Kim, C.-S.; Jeon, M.Y. Dynamic and static strain fiber Bragg grating sensor interrogation with a 1.3 μm Fourier domain mode-locked wavelength-swept laser. *Meas. Sci. Technol.* **2010**, *21*, 094008. [CrossRef]

17. Kwon, Y.S.; Ko, M.O.; Jung, M.S.; Park, I.G.; Kim, N.; Han, S.P.; Ryu, H.C.; Park, K.H.; Jeon, M.Y. Dynamic sensor interrogation using wavelength-swept laser with a polygon-scanner-based wavelength filter. *Sensors* **2013**, *13*, 9669–9678. [CrossRef]

18. Ko, M.O.; Kim, S.-J.; Kim, J.-H.; Lee, B.W.; Jeon, M.Y. Dynamic measurement for electric field sensor based on wavelength-swept laser. *Opt. Express* **2014**, *22*, 16139–16147. [CrossRef]

19. Ko, M.O.; Kim, S.-J.; Kim, J.-H.; Jeon, M.Y. In situ observation of dynamic pitch jumps of in-planar cholesteric liquid crystal layers based on wavelength-swept laser. *Opt. Express* **2018**, *26*, 28751–28762. [CrossRef]

20. Ahn, S.; Ko, M.O.; Kim, J.-H.; Chen, Z.; Jeon, M.Y. Characterization of second-order reflection bands from a cholesteric liquid crystal cell based on a wavelength-swept laser. *Sensors* **2020**, *20*, 4643. [CrossRef]

21. Lee, H.J.; Kim, S.-J.; Ko, M.O.; Kim, J.-H.; Jeon, M.Y. Tunable, multiwavelength-swept fiber laser based on nematic liquid crystal device for fiber-optic electric-field sensor. *Opt. Commun.* **2018**, *410*, 637–642. [CrossRef]

22. Park, J.; Kwon, Y.S.; Ko, M.O.; Jeon, M.Y. Dynamic fiber Bragg grating strain sensor interrogation with real-time measurement. *Opt. Fiber Technol.* **2017**, *38*, 147–153. [CrossRef]

23. Jeon, M.Y.; Kim, N.; Han, S.-P.; Ko, H.; Ryu, H.-C.; Yee, D.-S.; Park, K.H. Rapidly frequency-swept optical beat source for continuous wave terahertz generation. *Opt. Express* **2011**, *19*, 18364–18371. [CrossRef] [PubMed]

24. Han, G.-H.; Cho, S.-W.; Park, N.S.; Kim, C.-S. Electro-optic swept source based on AOTF for wavenumber-linear interferometric sensing and imaging. *Fibers* **2016**, *4*, 14. [CrossRef]

25. Park, N.S.; Chun, S.K.; Han, G.-H.; Kim, C.-S. Acousto-optic-based wavelength-comb-swept laser for extended displacement measurements. *Sensors* **2017**, *17*, 740. [CrossRef]

26. Chen, M.; Li, H.; Chen, R. Swept laser source based on acousto-optic tunable filter. *Proc. SPIE* **2014**, *9297*, 2071321. [CrossRef]

27. Srinivasan, V.J.; Huber, R.; Gorczynska, I.; Fujimoto, J.G.; Jiang, J.Y.; Reisen, P.; Cable, A.E. High-speed, high-resolution optical coherence tomography retinal imaging with a frequency-swept laser at 850 nm. *Opt. Lett.* **2007**, *32*, 361–363. [CrossRef]

28. Goda, K.; Fard, A.; Malik, O.; Fu, G.; Quach, A.; Jalali, B. High-throughput optical coherence tomography at 800 nm. *Opt. Express* **2012**, *20*, 19612–19617. [CrossRef]

29. Shirazi, M.F.; Jeon, M.; Kim, J. 850 nm centered wavelength-swept laser based on a wavelength selection galvo filter. *Chin. Opt. Lett.* **2016**, *14*, 011401. [CrossRef]

30. Lee, S.-W.; Song, H.-W.; Jung, M.-Y.; Kim, S.-H. Wide tuning range wavelength-swept laser with a single SOA at 1020 nm for ultrahigh resolution Fourier-domain optical coherence tomography. *Opt. Express* **2011**, *19*, 21227–21237. [CrossRef]

31. Shramenko, M.V.; Chamorovskiy, A.; Lyu, H.C.; Lobintsov, A.A.; Karnowski, K.; Yakubovich, S.D.; Wojtkowski, M. Tunable semiconductor laser at 1025–1095 nm range for OCT applications with an extended imaging depth. *Proc. SPIE* **2015**, *9312*, 93123B. [CrossRef]

32. Cao, J.; Wang, P.; Zhang, Y.; Shi, G.; Wu, B.; Zhang, S.; Liu, Y. Methods to improve the performance of the swept source at 1.0 μm based on a polygon scanner. *Photonics Res.* **2017**, *5*, 245–250. [CrossRef]

33. Wang, L.; Wan, M.; Shen, Z.; Wang, X.; Cao, Y.; Feng, X.; Guan, B. Wavelength-swept fiber laser based on bidirectional used linear chirped fiber Bragg grating. *Photonics Res.* **2017**, *5*, 219–223. [CrossRef]

34. Cao, J.; Wang, L.; Lu, Z.; Wang, G.; Wang, X.; Ran, Y.; Feng, X.; Guan, B. High-speed refractive index sensing system based on Fourier domain mode locked laser. *Opt. Express* **2019**, *27*, 7988–7996. [CrossRef]

35. Sung, J.-Y.; Chen, J.-K.; Liaw, S.-K.; Kishikawa, H.; Goto, N. Fiber Bragg grating sensing system with wavelength-swept laser distribution and self-synchronization. *Opt. Lett.* **2020**, *45*, 5436–5439. [CrossRef]

36. Tokurakawa, M.; Daniel, J.M.O.; Chenug, C.S.; Liang, H.; Clarkson, W.A. Wavelength-swept Tm-doped fiber laser operating in the two-micron wavelength band. *Opt. Express* **2014**, *22*, 20014–20019. [CrossRef]

37. Tokurakawa, M.; Daniel, J.M.O.; Chenug, C.S.; Liang, H.; Clarkson, W.A. Ultra-broadband wavelength-swept Tm-doped fiber laser using wavelength-combined gain stages. *Opt. Express* **2015**, *23*, 471–476. [CrossRef]

38. Tan, S.; Yang, L.; Wei, X.; Li, C.; Chen, N.; Tsia, K.K.; Wong, K.K.Y. High-speed wavelength-swept source at 2.0 μm and its application in imaging through a scattering medium. *Opt. Lett.* **2017**, *42*, 1540–1543. [CrossRef]

39. Xu, J.; Zhang, C.; Xu, J.; Wong, K.K.Y.; Tsia, K.K. Megahertz all-optical swept-source optical coherence tomography based on broadband amplified optical time-stretch. *Opt. Lett.* **2014**, *39*, 622–625. [CrossRef]
40. Wei, X.; Lay, A.K.S.; Xu, Y.; Tsia, K.K.; Wong, K.K.Y. 28 MHz swept source at 1.0 μm for ultrafast quantitative phase imaging. *Biomed. Opt. Express* **2015**, *6*, 3855–3864. [CrossRef]
41. Huang, D.; Li, F.; Shang, C.; Cheng, Z.; Wai, P.K.A. Reconfigurable time-stretched swept laser source with up to 100 MHz sweep rate, 100 nm bandwidth, and 100 mm OCT imaging range. *Photonics Res.* **2020**, *8*, 1360–1367. [CrossRef]
42. Huang, D.; Li, F.; He, Z.; Cheng, Z.; Shang, C.; Wai, P.K.A. 400 MHz ultrafast optical coherence tomography. *Opt. Lett.* **2020**, *45*, 6675–6678. [CrossRef]
43. Li, B.; Zhang, C.; Kang, J.; Wei, X.; Tan, S.; Wong, K.K.Y. 109 MHz optical tomography using temporal magnification. *Opt. Lett.* **2015**, *40*, 2965–2968. [CrossRef]
44. Kang, J.; Feng, P.; Wei, X.; Lam, E.Y.; Tsia, K.K.; Wong, K.K.Y. 102-nm, 44.5-MHz inertial-free swept source by mode-locked fiber laser and time stretch technique for optical coherence tomography. *Opt. Express* **2018**, *26*, 4370–4381. [CrossRef]
45. Gerguis, J.O.; Sabry, Y.M.; Omran, H.; Khalil, D. Spectroscopic Gas Sensing Based on a MEMS-SOA Swept Fiber Laser Source. *J. Light. Technol.* **2019**, *37*, 5354–5360. [CrossRef]
46. Tsai, T.H.; Potsaid, B.; Tao, Y.K.; Jayaraman, V.; Jiang, J.; Heim, P.J.S.; Kraus, M.F.; Zhou, C.; Hornegger, J.; Mashimo, H.; et al. Ultrahigh speed endoscopic optical coherence tomography using micromotor imaging catheter and VCSEL technology. *Biomed. Opt. Express* **2013**, *4*, 1119. [CrossRef]
47. Yamashita, S.; Asano, M. Wide and fast wavelength-tunable mode-locked fiber laser based on dispersion tuning. *Opt. Express* **2006**, *14*, 9299–9306. [CrossRef]
48. Takubo, Y.; Yamashita, S. High-speed dispersion-tuned wavelength-swept fiber laser using a reflective SOA and a chirped FBG. *Opt. Express* **2013**, *21*, 5130–5139. [CrossRef]
49. Takubo, Y.; Shirahata, T.; Yamashita, S. Optinization of a dispersion-tuned wavelength-swept fiber laser for optical coherence tomography. *Appl. Opt.* **2016**, *55*, 7749–7755. [CrossRef]
50. Cao, J.; Wang, P.; Zhang, Y.; Shi, G.; Liu, Y. Experimental and theoretical investigation of polygon-based swept source with continuous adjustable free spectral range. *Opt. Commun.* **2021**, *478*, 126401. [CrossRef]

Letter

Characterization of Second-Order Reflection Bands from a Cholesteric Liquid Crystal Cell Based on a Wavelength-Swept Laser

Soyeon Ahn [1,†], Myeong Ock Ko [2,†], Jong-Hyun Kim [1,3], Zhongping Chen [4] and Min Yong Jeon [1,3,*] 

[1] Department of Physics, Chungnam National University, 99 Daehak-ro Yuseong-gu, Daejeon 34134, Korea; ahnsoyen5@naver.com (S.A.); jxk97@cnu.ac.kr (J.-H.K.)

[2] Core Technology R&D Team, Samsung Electronics, Hwaseong-si, Gyeonggi-do 18448, Korea; myeongock08@naver.com

[3] Instituted of Quantum Systems (IQS), Chungnam National University, 99 Daehak-ro Yuseong-gu, Daejeon 34134, Korea

[4] Beckman Laser Institute, UC Irvine, Irvine, CA 92612, USA; z2chen@uci.edu

* Correspondence: myjeon@cnu.ac.kr; Tel.: +82-42-821-5459

† These authors contributed equally in this work.

Received: 30 June 2020; Accepted: 15 August 2020; Published: 18 August 2020

Abstract: We report the results of an experimental study of the characterization of second-order reflection bands from a cholesteric liquid crystal (CLC) cell that depends on the applied electric field, using a wide bandwidth wavelength-swept laser. The second-order reflection bands around 1300 nm and 1500 nm were observed using an optical spectrum analyzer when an electric field was applied to a horizontally oriented electrode cell with a pitch of 1.77 μm. A second-order reflection spectrum began to appear when the intensity of the electric field was 1.03 $V_{rms}/\mu m$ with the angle of incidence to the CLC cell fixed at 36°. The reflectance increased as the intensity of the electric field increased at an angle of incidence of 20°, whereas at an incident angle of 36°, when an electric field of a predetermined value or more was applied to the CLC cell, it was confirmed that deformation was completely formed in the liquid crystal and the reflectance was saturated to a constant level. As the intensity of the electric field increased further, the reflection band shifted to a longer wavelength and discontinuous wavelength shift due to the pitch jump was observed rather than a continuous wavelength increase. In addition, the reflection band changed when the angle of incidence on the CLC cell was changed. As the angle of incidence gradually increased, the center wavelength of the reflection band moved towards shorter wavelengths. In the future, we intend to develop a device for optical wavelength filters based on side-polished optical fibers. This is expected to have a potential application as a wavelength notch filter or a bandpass filter.

Keywords: fiber laser; wavelength-swept laser; cholesteric liquid crystal; bandpass filter

---

1. Introduction

Cholesteric liquid crystals (CLCs) or chiral nematic liquid crystals (NLCs) are liquid crystals (LCs) with chiral dopants that induce a periodic helical structure. They exhibit a helical structure in which the directors of the LCs are twisted and arranged in layers along the spiral axis. The details of the optical properties of cholesteric liquid crystal according to the applied electric field have been discussed for a long time [1–3]. The distance by which the director of the LC is rotated 360° in the axial direction of the helical is called a pitch. When the polarization direction of the incident light has the same handedness of the helical structure and the periodicity of the pitch satisfies the Bragg condition, the incident lights have selective reflection characteristics. For normal incidence, the reflection wavelength λ_o is given by [4–6]:

$$\lambda = \lambda_o = \bar{n}P \tag{1}$$

where $\bar{n}$ is the average of the ordinary (n_o) and extraordinary (n_e) refractive indices of the NLC and P is the pitch of the CLC. When a broadband light source is incident to the CLC cell, the reflection band $\Delta\lambda$ can be formed when the Bragg condition is satisfied as follows:

$$\Delta\lambda = \Delta nP = (n_e - n_o)P \tag{2}$$

where $\Delta n = (n_e - n_o)$ is the birefringence. The periodic helical structure of the CLC cell can be changed by various external stimuli such as heat, electric fields or magnetic fields [7–9]. Due to these characteristics, CLCs have been studied regarding the fabrication of optical devices for various applications such as liquid crystal displays (LCDs), dye lasers, notch filters, optical sensors and mirrors [10–23]. However, when the CLC cell is inclined with respect to the incident beam the reflection band shifts to a shorter wavelength due to the Bragg condition and a higher order reflection band appears [12,24]. For an inclined CLC cell, the center wavelength due to the Bragg reflection is given by:

$$m\lambda = \bar{n}P \cos\Theta \tag{3}$$

$$\Theta = \sin^{-1}\left(\frac{1}{\bar{n}}\sin\theta\right) \tag{4}$$

where θ is the angle of the incidence of the beam in the direction normal to the CLC cell and Θ is the angle between the direction of propagation and helical axis [25]. Higher order reflection bands have a much narrower bandwidth than the first-order reflection band so this phenomenon is useful for fabricating optical bandpass filters [12]. When an electric field is applied perpendicular to the helical axis of an LC having positive anisotropy, the deformation occurs because the direction of the LC is aligned in the field direction and thus the pitch increases as the deformation increases. It is known that the higher order reflection band originates from the non-sinusoidal distribution of a refractive index caused by the in-plane-field-induced distortion of the CLC helices [12]. In particular, in the case of CLCs, an electric field or magnetic field applied perpendicular to the helical axis produces a non-uniform twist of the helical structure and increases birefringence of the CLC. Chou et al. calculated the wave equation to determine the transmission coefficient of the CLC and when an electric field was applied, a second reflection appeared, which explains why the birefringence increased [26]. In addition, Dumitrascu et al. reported that when the electric field was applied perpendicular to the helical axis, the ordinary refractive index n_o was hardly affected by the electric field but the extraordinary refractive index n_e increased as the intensity of the electric field increased. In addition, saturation occurred in reflectance above some electric fields [27]. Therefore, it can be explained that the increase in reflectance is due to the increase in the birefringence of the LC when an electric field is applied [12,26,27].

A broadband light source is required to measure the wide reflection band in the infrared region. Wavelength swept lasers (WSLs) in the 1300 nm band or 1500 nm band have a wide scanning bandwidth of ~100 nm or wider, which is suitable for measuring these characteristics [28–30]. Moreover, by combining the two wavelength bands, it is possible to observe the reflection band in a much wider band region. Furthermore, it is a light source that is easy to observe regarding dynamic variations in reflection bands, with respect to changes in the electric field intensity [30].

In this paper, we successfully observed second-order reflection bands from a CLC cell that varied dependent on the applied electric field using a wide bandwidth WSL. As the intensity of the electric field was increased further, the reflectance of the CLC cell also increased. In addition, changes in the transmission spectrum were observed in response to changing the angle of the beam incident on the CLC cell under the application of a constant electric field.

2. Fabrication of the CLC Cell

In the experiments, an NLC E7 (Qingdao QY liquid crystal) and a chiral dopant R811 were mixed to produce a right-handed CLC. The ordinary and extraordinary refractive indices of the CLC were 1.5014 and 1.6885, respectively. Figure 1 shows the fabrication process of the CLC cell. First of all, the cell was prepared by cutting the slide glass to a size of ~13 mm × 18 mm using a diamond knife. The cells were then washed for 15 min in an ultrasonic cleaner in the order of de-ionized water, acetone and ethanol. Figure 1a shows an electrode substrate coated with 8 nm and 20 nm of Ti and Au, respectively, by masking the optical fiber with Kapton tape on the cleaned glass and then using an electron beam (e-beam) vacuum evaporator. A thin coating of Ti before Au on the slide glass substrate helped the Au adhere well to it. The gap between the two in-plane electrodes was ~400 μm. The electrode substrate was then washed and dried again. The polyimide solution of AL-3046 was spread onto the substrates using a pipette; it was then spin-coated at approximately 3000 rpm for 30 s (Figure 1b) and then baked for 1 h at 180 °C on a hot plate (Figure 1c). The substrate was then rubbed ~20 times in a specific direction using a rubbing machine (Figure 1d). The rubbing machine used in this experiment was homemade and the substrate was rubbed with a velvet rubbing fabric. When rubbing the electrode substrate, it was necessary to rub in a direction parallel to the electrode. This process allowed the LC molecules to align in the rubbing direction. The rubbing direction of the two substrates was then anti-parallel and the gap between the two substrates was created using a 20 μm film spacer and epoxy. In the case of an electrode cell, one substrate became an electrode substrate and the other became a general glass substrate. Figure 1e shows an indium solder wire attached to an electrode cell. This ensured that the electric field was applied perpendicular to the helical axis. In this case, the LC was injected after indium soldering because indium soldering can affect the LC by dissolving the indium at a high temperature. In the case of E7, it is a liquid crystal phase at room temperature and becomes isotropic phase when it is about 60 °C. Therefore, the CLC cell was placed on a hot plate and mixed with a chiral dopant in an isotropic state. After that, the CLC was mixed well using a vortex mixer or stirrer and then injected between two substrates using a pipette.

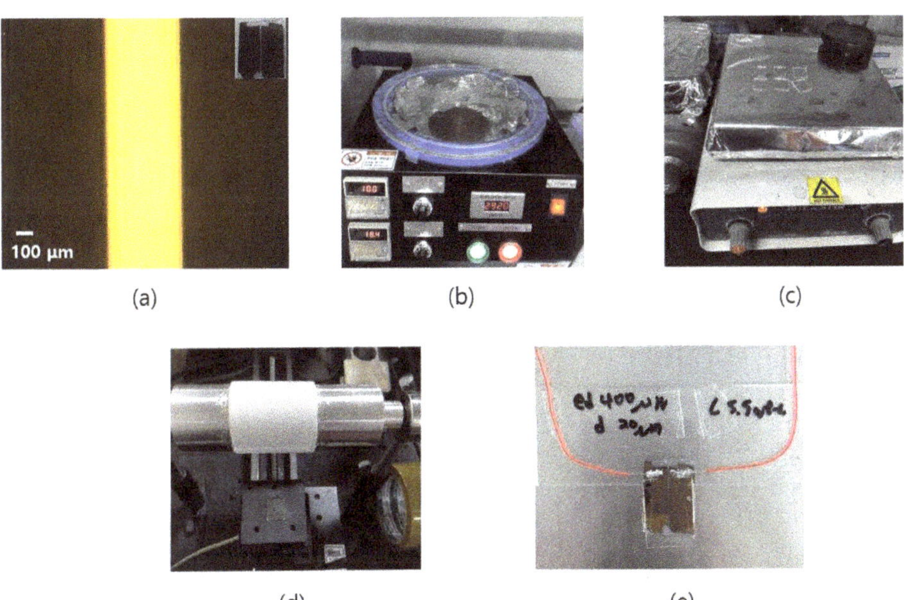

Figure 1. Fabrication process of the CLC cell; (**a**) electrode substrates, (**b**) spin coating, (**c**) baking, (**d**) rubbing and (**e**) the fabricated CLC cell.

Figure 2 shows the fabricated CLC cell structure. It consisted of glass substrates, polyimide layers, electrode layers and a CLC layer. Two flat electrodes provided the in-plane electric field. The gap between the two electrodes was ~400 μm. The thickness of Au and Ti as the electrode layer together was ~28 nm and the CLC cell thickness was ~20 μm. A homogeneously aligned CLC cell was driven by an electric field perpendicular to the helical axis. Since the distance between the in-plane electrodes was sufficiently wide compared with the cell gap, it can be considered that the helical structure cell was subjected to a uniform perpendicular electric field. Since the pitch of the CLC cell changed according to the intensity of the in-plane electric field, the reflection band of the CLC cell also changed by the applied electric field [30].

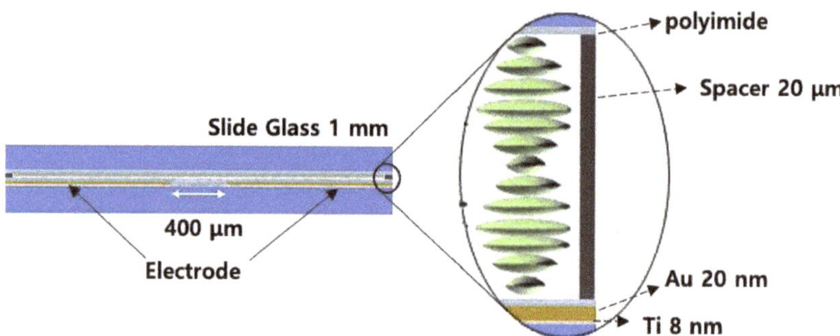

Figure 2. Structure of the CLC cell.

The chiral dopant concentration of the CLC cell was 5.5 wt%. In order to measure the pitch of the CLC cell, two methods were used: one method was to measure using the Cano wedge cell having a constant slope and the other method was to measure using the transmittance spectrum from the CLC cell. The Cano wedge cell had dislocation defect walls at half-pitch because the molecules discontinuously arranged when the thickness changed by half-pitch. The pitch can be measured by the slope of the wedge cell and the distance between the dislocation lines [31]. The measurement was performed four times by the wedge cell method and the measured pitch value was taken as the average value of four measurements. Figure 3 shows the photograph of the texture for the Cano wedge cell when a CLC with a concentration of 5.5 wt% was injected into a wedge cell. The color change in the wedge cell was due to the inclination of the cell, which is a phenomenon that appears due to the difference in the cell gap. The dislocation lines were formed on the texture of the CLC Cano wedge cell.

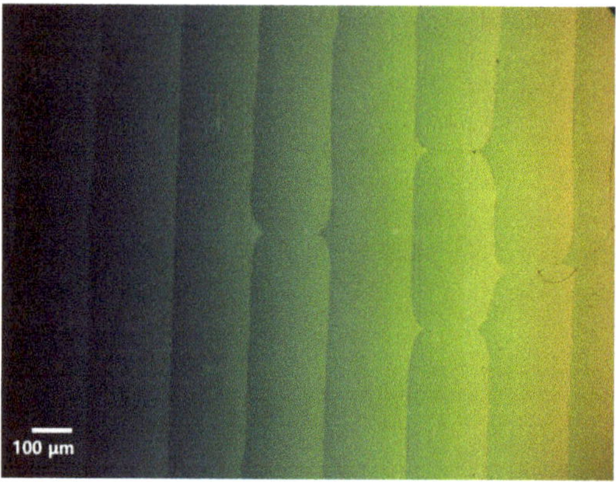

Figure 3. Photograph of the CLC texture.

The next method was to estimate the pitch of the CLC cell using the transmittance spectrum. The WSL was positioned incident to the normal direction on the cell. The electric field applied to the CLC cell was fixed to 0.49 $V_{rms}/\mu m$. The second-order reflection spectrum was achieved from 1331 nm to 1368 nm using the transmittance spectrum as shown in Figure 4. The edge-to-edge bandwidth of the reflection band was ~37 nm. The pitch was calculated to be 1.77 μm using Equation (3).

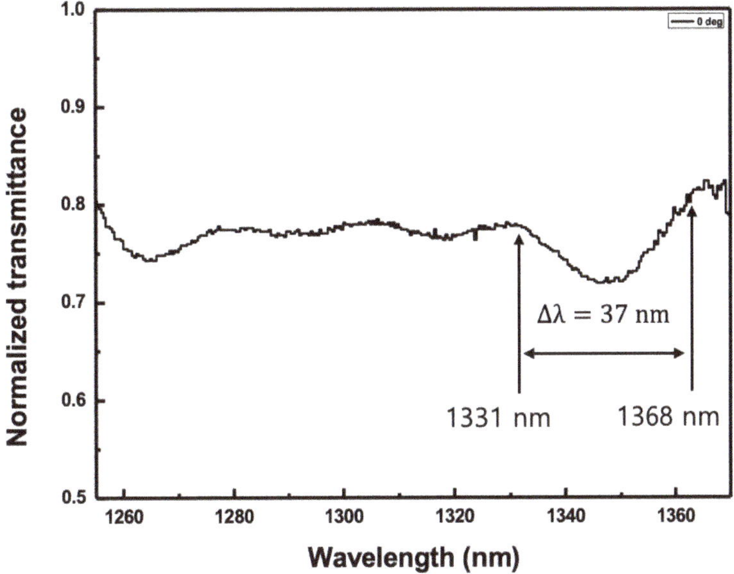

Figure 4. Second-order reflection spectrum when the light was incident to the normal direction of the CLC cell when the electric field applied to the CLC cell was fixed to 0.49 $V_{rms}/\mu m$.

3. Experiments

The measurements of the optical characteristics of the CLC cell were carried out using a broadband WSL. With WSL light sources, wavelength characteristics can be observed over a wider range than using a semiconductor optical amplifier (SOA). Figure 5a shows the experimental setup used to measure the second-order reflection spectra of the CLC cell using two WSLs. Two WSLs were combined to form a broadband WSL from 1300 nm to 1500 nm bands. The first WSL, consisting of SOA1 and SOA2, operated in the 1300 nm band. The second WSL, consisting of SOA3 and SOA4, operated in the 1500 nm band. Each WSL consisted of two SOAs, an optical isolator, three polarization controllers, an optical output coupler, an optical circulator, a diffraction grating with 600 grooves/mm, two achromatic doublet lenses and a polygonal wavelength scanning filter [30]. The 10-dB scanning bandwidths of the WSLs around 1300 nm and 1500 nm were ~118.4 nm and ~116.8 nm, respectively, as shown in Figure 5b. The scanning range around 1300 nm and 1500 nm were from 1253.2 nm to 1371.6 nm and 1470 nm to 1586.8 nm, respectively. The scanning rate and the average output power of the WSL were 3.6 kHz and ~13 dBm, respectively. The dotted box in Figure 5a shows the measurement setup used to measure the transmittance band spectra of the CLC cell.

The output from the WSL was incident to the measurement setup. The beam was set to be right-handed circularly polarized through a polarization beam splitter (PBS) and a quarter wave plate (QWP). It was then positioned to be incident vertically on the CLC cell. The CLC cell was positioned in the depth of the focus (DOF) of the beam. The size of the beam was ~85 μm within the area of the range between the electrodes. Since the distance between the electrodes was far enough for the gap size of the CLC cell, it can be assumed that when an in-plane electric field is applied, a uniform electric field is applied to the central area of the CLC cell. The beam transmitted through the CLC cell was measured using an optical spectrum analyzer (OSA). A 5 kHz sinusoidal wave with alternating current (AC) voltage was applied to the CLC cell parallel to the surface using a function generator (Agilent) and an amplifier (Trek). The transmitted beam was measured according to the applied electric field or the angle of incidence of the CLC cell using the OSA.

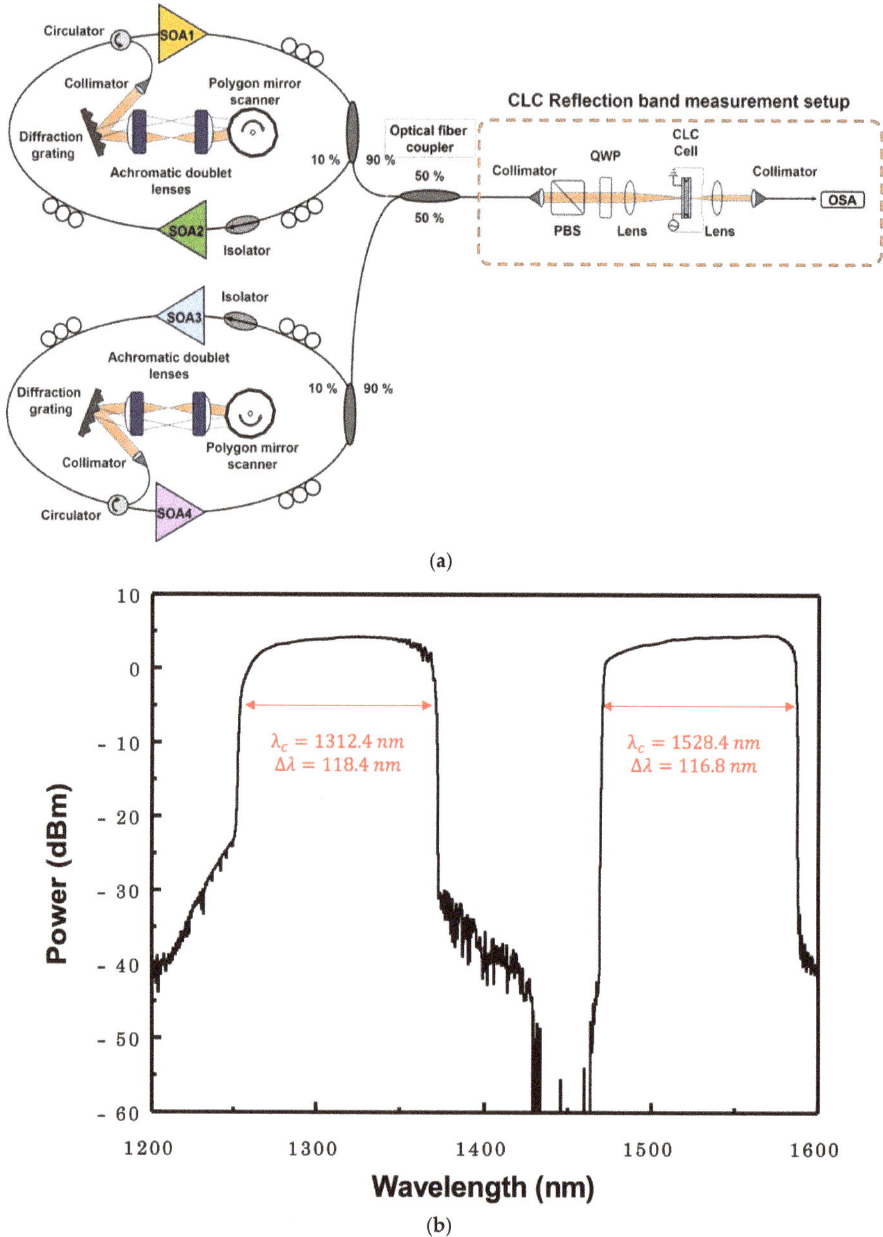

Figure 5. (a) Experimental setup for measuring the CLC reflection band and (b) optical spectrum output from the WSL.

The voltage from 50 mV$_{rms}$ to 270 mV$_{rms}$ was applied to the CLC cell through the amplifier; the voltage was increased in 10 mV$_{rms}$ steps and the corresponding electric field ranged from 0.18 V$_{rms}$/μm to 1.03 V$_{rms}$/μm. Figure 6 shows the normalized transmitted spectra according to the applied electric field when the angle of incidence to the CLC cell was fixed to 20°. The normalized transmitted spectra can be achieved with the following steps. First of all, in the measurement setup, the transmission spectrum was measured without the CLC cell. It can be used as a reference spectrum. After fixing the cell on the rotation stage, it was placed in the DOF region and the spectrum according to the electric field intensity or the angle of incidence was measured. The normalized transmission spectrum was obtained by differentiating this value from the reference spectrum. At the electric field of 0.36 V$_{rms}$/μm, the second-order reflection band of the CLC cell began to appear around 1314 nm of the reflection band as shown in Figure 6a. When the electric field increased up to 0.77 V$_{rms}$/μm, there was no change in pitch, meaning that the reflection band was almost maintained. The relative reflectance in the case of Figure 6a was measured to be ~0.3 or less. However, when it increased above 0.85 V$_{rms}$/μm, a pitch jump occurred and the center wavelength of the reflection band moved to near 1350 nm as shown in Figure 6b. The second reflection band was maintained until the electric field was 0.99 V$_{rms}$/μm. It was shown that the relative reflectance increased as the voltage increased. The relative reflectance in the case of Figure 6b was measured to be ~0.45 or less. When the angle of incidence was 20°, the deformation was formed as the electric field was increased so the relative reflectance tended to increase. The increase in reflectance was caused by the increase in the birefringence of the CLC when an electric field was applied [26,27].

Next, the angle of incidence was changed to 36° to observe the second-order reflection characteristics according to the applied electric field. The voltage from 270 mV$_{rms}$ to 360 mV$_{rms}$ was applied to the CLC cell; the voltage was increased in 10 mV$_{rms}$ steps and the corresponding electric field ranged from 1.03 V$_{rms}$/μm to 1.39 V$_{rms}$/μm. Figure 7 shows the normalized transmitted spectra according to the applied electric field when the angle of incidence to the CLC cell was fixed to 36°. At the electric field of 1.03 V$_{rms}$/μm, the second-order reflection band of the CLC cell began to appear around 1274 nm of the reflection band. When the electric field increased from 1.07 to 1.11 V$_{rms}$/μm, there was no change in pitch. In these cases, most reflection bands had a width of ~25 nm. However, when it increased above 1.15 V$_{rms}$/μm, a pitch jump occurred and the center wavelength of the reflection band moved to near 1313 nm. The second reflection band was maintained until the electric field was 1.23 V$_{rms}$/μm. In these cases, most reflection bands have a width of ~30 nm. They were observed slightly wider than in the previous cases. As the intensity of the electric field increased to 1.27 V$_{rms}$/μm, the reflection band shifted to a longer wavelength of 1349 nm. In addition, it was found that the band width was further increased to 36 nm or more. In an electric field higher than 1.39 V$_{rms}$/μm, a pitch jump occurred and it could not be measured any more beyond the scanning wavelength range of the WSL. This discontinuous change in the reflection band indicated that the pitch of the CLC cell discontinuously varied as the intensity of the electric field increased. The discretization of the pitch according to the intensity of the electric field in the LC was strongly anchored at the surface boundary so the pitch increased discontinuously when the electric field increased to a certain value or more [12,30,32].

In order to observe the second-order reflectance over a 1500 nm band, a higher electric field was applied to the CLC cell. Figure 8 shows the reflected spectra according to the applied electric field of more than 1.54 V$_{rms}$/μm when the angle of incidence to the CLC cell was fixed to 36°. The full reflected band structure could not be observed because the width of the reflection bands was outside the measurable wavelength range. However, as shown in Figure 8, the short edge of the reflection band was measured when the applied electric field was more than 1.54 V$_{rms}$/μm. When the applied electric field was 1.54 V$_{rms}$/μm, the short edge of the reflection band appeared near 1515 nm. On the other hand, when the applied electric field was 1.6 V$_{rms}$/μm, the short edge of the reflection band appeared near 1565 nm. However, when the applied electric field was 1.64 V$_{rms}$/μm, the short edge of the reflection band could not be observed because it was outside the range of WSL.

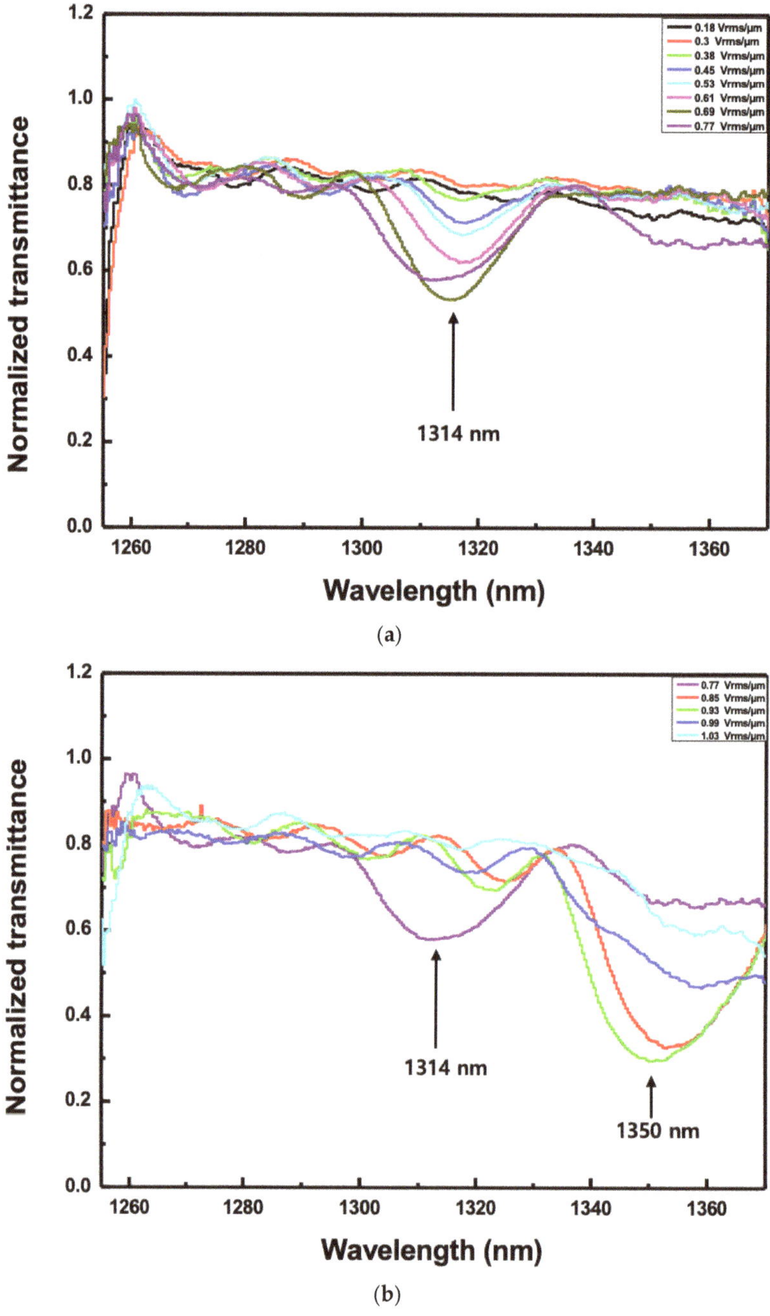

Figure 6. The normalized transmitted spectra according to the applied electric field when the angle of incidence to the CLC cell was fixed to 20°. (**a**) The electric field ranged from 0.18 V_{rms}/µm to 0.77 V_{rms}/µm and (**b**) the electric field ranged from 0.77 V_{rms}/µm to 1.03 V_{rms}/µm.

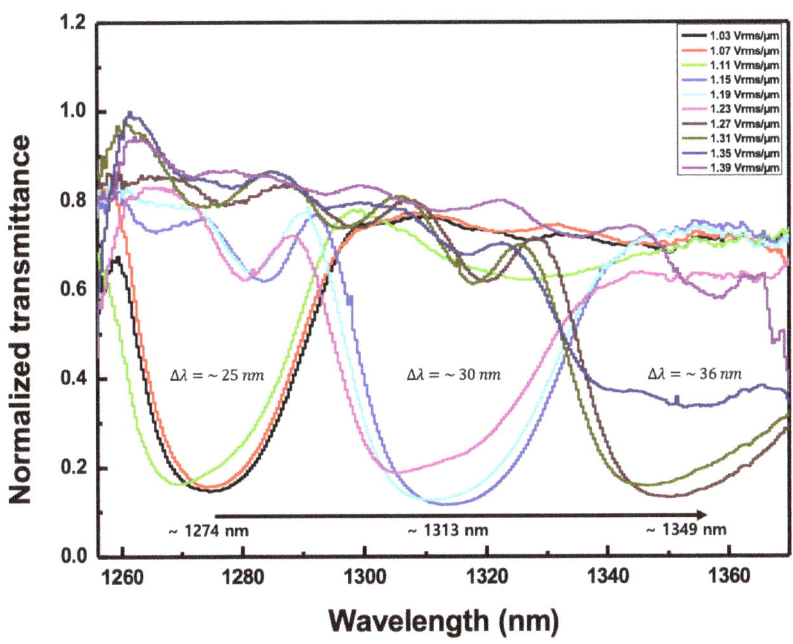

Figure 7. The normalized transmitted spectra according to the applied electric field when the angle of incidence to the CLC cell was fixed to 36°.

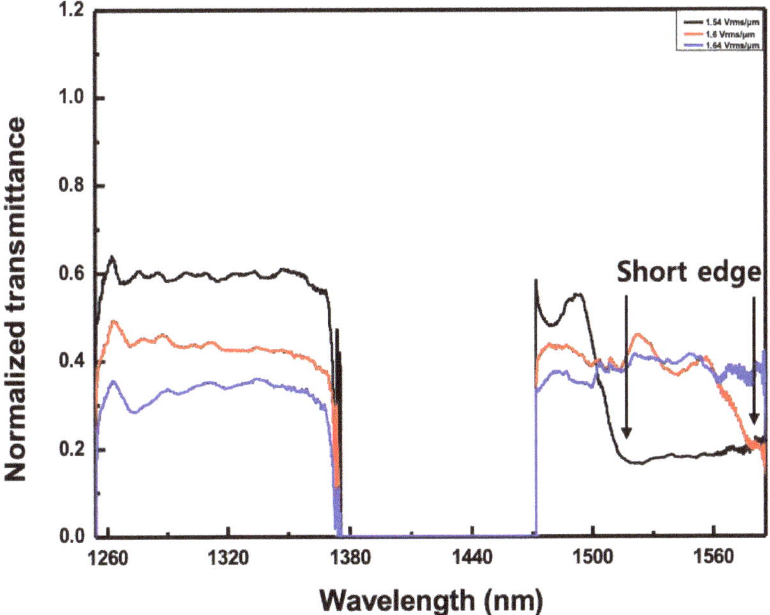

Figure 8. The normalized transmitted spectra according to the applied electric field more than 1.54 V_{rms}/μm when the angle of incidence to the CLC cell was fixed to 36°.

Figure 9 shows the plots of the relative reflectance versus the electric field applied to the CLC cell for the reflection bands in Figure 7 when the angle of incidence to the CLC cell was fixed to 36°. The relative reflectance was obtained at values of ~0.6 as shown in Figure 8. When the angle of incidence was 20°, the relative reflectance tended to increase but at 36°, the relative reflectance was somewhat constant. This is the reason that when the intensity of the applied electric field increased above a certain value, the deformation was completely formed and thus the reflectance was saturated [27].

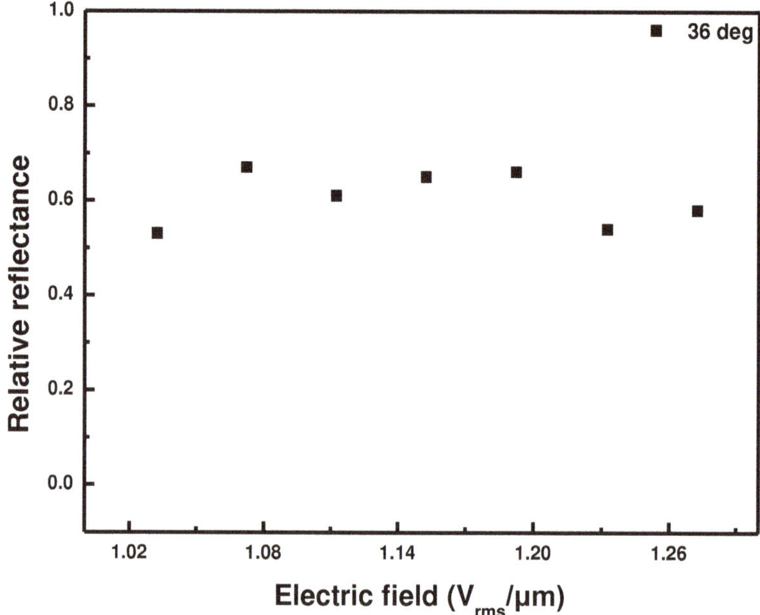

Figure 9. Relative reflectance of the CLC cell according to the applied electric field when the angle of incidence to the CLC cell was fixed to 36°.

Figure 10a shows the normalized transmitted spectra according to the angle of incidence to the CLC cell. The electric field applied to the CLC cell was fixed at 0.49 V$_{rms}$/μm. The transmitted spectra were measured while inclining the CLC cell at an interval of 2° from 0° to 26° with respect to the incident light. The center wavelength of the reflection band when the beam was incident vertically on the CLC cell was 1348 nm. As the angle of incidence was gradually increased, the center wavelength of the reflection band moved towards shorter wavelengths. When the angle of incidence was 12°, the center wavelength of the reflection band was 1331 nm. When it was more inclined at 20°, the reflection band moved to 1312 nm and when the angle of incidence was 26°, it moved to 1294 nm. These phenomena happened because the pitch of the CLC was different depending on the angle of incidence of the beam with respect to the CLC cell. Figure 10b shows the variation of the short edge wavelength according to the angle of incidence of the beam on the CLC cell. As the angle of inclination of the CLC cell increased, the short edge wavelength moved towards shorter wavelengths, as shown in Figure 10. The experimental data almost coincided to the value calculated using Equation (3).

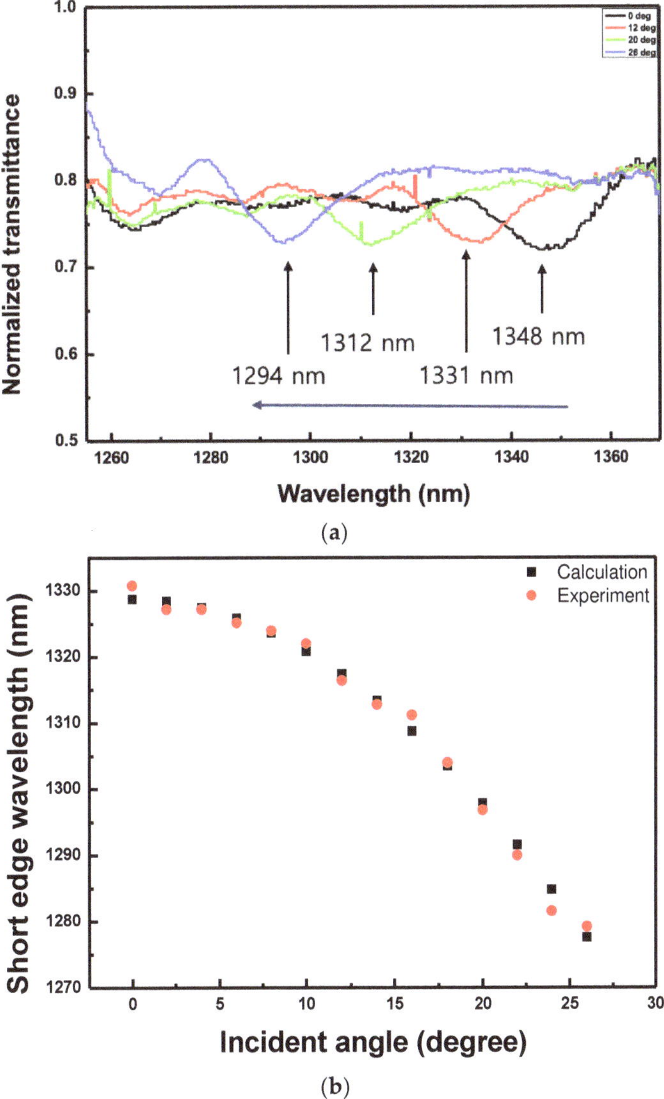

Figure 10. (**a**) Normalized transmitted spectra and (**b**) variation of the short edge wavelength according to the angles of incidence of the beam on the CLC cell when the electric field applied to the CLC cell was fixed to 0.49 $V_{rms}/\mu m$.

4. Conclusions

We have successfully investigated second-order reflection bands from a cholesteric liquid crystal (CLC) cell varied dependent on the applied electric field and on the angle of incidence of the beam on the CLC cell. In order to observe the reflection spectrum, two wide-band wavelength-swept lasers (1300 nm and 1500 nm band) were used as an optical source. Second-order reflection spectra were observed using an optical spectrum analyzer after applying an electric field to a horizontally oriented electrode cell with a pitch of 1.77 μm. Second-order reflection spectra began to appear when the intensity of the electric field was 1.03 $V_{rms}/\mu m$ with the angle of incidence to the CLC cell fixed at 36°.

As the intensity of the electric field was increased further, the reflection band shifted discontinuously towards longer wavelengths. Most of the reflection bands were about 25 nm to 36 nm, which confirmed the possibility as a wavelength bandpass filter and confirmed the possibility as an electric field sensor by using a wavelength change according to the intensity of the electric field. In addition, the reflection band changed when the angle of incidence on the CLC cell was changed under a fixed electric field. As the angle of incidence was gradually increased, the center wavelength of the reflection band moved towards a shorter wavelength. In the future, we intend to develop a device for optical wavelength filters based on side polished optical fibers. This is expected to have a potential application as a wavelength notch filter or a bandpass filter.

Author Contributions: S.A. and M.O.K. conceived and designed the experiments; J.-H.K. reviewed and analyzed the data; S.A. and M.O.K. contributed to writing—original draft preparation; Z.C. and M.Y.J. writing—review and editing; M.Y.J. supervised. All authors have read and agreed to the published version of the manuscript.

Funding: This research was supported by Chungnam National University of Korea and was supported by a Basic Science Research Program through the National Research Foundation of Korea (NRF) funded by the Ministry of Science, ICT and future Planning (NRF-2017R1A2B4008212, NRF-2019R1A2C1084933, NRF-2020R1A6A1A03047771) and was supported by a Korea Institute for Advancement of Technology (KIAT) grant funded by the Korea Government (MOTIE) (P0008458, The Competency Development Program for Industry Specialist).

Conflicts of Interest: The authors declare no conflict of interest.

References

1. De Gennes, P.-G.; Prost, J. *The Physics of Liquid Crystals*, 2nd ed.; Oxford University Press: Oxford, UK, 1993.
2. Dunmur, D.; Fukuda, A.; Luckhurst, G.R. *Physical Properties of Liquid Crystals. Nematics (EMIS Datareviews Series No. 25)*; INSPEC: London, UK, 2001.
3. Goodby, J.W. Chirality in liquid crystals. *J. Mater. Chem.* **1991**, *1*, 307–318. [CrossRef]
4. Xianyu, H.; Faris, S.; Crawford, G.P. In-plane switching of cholesteric liquid crystals for visible and near-infrared applications. *Appl. Opt.* **2004**, *43*, 5006–5015. [CrossRef] [PubMed]
5. Mitov, M.; Dessaud, N. Going beyond the reflectance limit of cholesteric liquid crystals. *Nat. Mater.* **2006**, *5*, 361–364. [CrossRef] [PubMed]
6. Zografopoulos, D.C.; Kriezis, E.E.; Mitov, M.; Binet, C. Theoretical and experimental optical studies of cholesteric liquid crystal films with thermally induced pitch gradients. *Phys. Rev. E* **2006**, *73*. [CrossRef] [PubMed]
7. Yang, D.K.; West, J.L.; Chien, L.C.; Doane, J.W. Control of reflectivity and bistability in display using cholesteric liquid crystals. *J. Appl. Phys.* **1994**, *76*, 1331–1333. [CrossRef]
8. Coles, H.; Demus, D.; Goodby, J.; Gray, G.W.; Spiess, H.W.; Vill, V. Chiral Nematics: Physical properties and applications. In *Handbook of Liquid Crystals Set*; Wiley-VCH Verlag GmbH: Weinheim, Germany, 2008; pp. 335–409. [CrossRef]
9. Hrozhyk, U.A.; Serak, S.V.; Tabiryan, N.V.; White, T.J.; Bunning, T.J. Optically switchable, rapidly relaxing cholesteric liquid crystal reflectors. *Opt. Express* **2010**, *18*, 9651–9657. [CrossRef] [PubMed]
10. Zhu, X.; Ge, Z.; Wu, T.X.; Wu, S.-T. Transflective liquid crystal displays. *J. Disp. Technol.* **2005**, *1*, 15–29. [CrossRef]
11. Schmidtke, J.; Jünnemann, G.; Keuker-Baumann, S.; Kitzerow, H.S. Electrical fine tuning of liquid crystal lasers. *Appl. Phys. Lett.* **2012**, *101*, 051117. [CrossRef]
12. Rumi, M.; White, T.J.; Bunning, T.J. Reflection spectra of distorted cholesteric liquid crystal structures in cells with interdigitated electrodes. *Opt. Express* **2014**, *22*, 16510–16519. [CrossRef] [PubMed]
13. Palto, S.P.; Barnik, M.I.; Geivandov, A.R.; Kasyanova, I.V.; Palto, V.S. Spectral and polarization structure of field-induced photonic bands in cholesteric liquid crystals. *Phys. Rev. E* **2015**, *92*. [CrossRef] [PubMed]
14. McConney, M.E.; Tondiglia, V.P.; Natarajan, L.V.; Lee, K.M.; White, T.J.; Bunning, T.J. Electrically induced color changes in polymer-stabilized cholesteric liquid crystals. *Adv. Opt. Mater.* **2013**, *1*, 417–421. [CrossRef]
15. Xiang, J.; Li, Y.; Li, Q.; Paterson, D.A.; Storey, J.M.D.; Imrie, C.T.; Lavrentovich, O.D. Electrically tunable selective reflection of light from ultraviolet to visible and infrared by heliconical cholesterics. *Adv. Mater.* **2015**, *27*, 3014–3018. [CrossRef] [PubMed]

16. Petriashvili, G.; Japaridze, K.; Devadze, L.; Zurabishvili, C.; Sepashvili, N.; Ponjavidze, N.; De Santo, M.P.; Matranga, M.A.; Hamdi, R.; Ciuchi, F.; et al. Paper like cholesteric interferential mirror. *Opt. Express* **2013**, *21*, 20821–20830. [CrossRef] [PubMed]

17. Shibaev, P.V.; Schlesier, C. Distant mechanical sensors based on cholesteric liquid crystals. *Appl. Phys. Lett.* **2012**, *101*, 193503. [CrossRef]

18. Mitov, M. Cholesteric liquid crystals with a broad light reflection band. *Adv. Mater.* **2012**, *24*, 6260–6276. [CrossRef]

19. Hennig, G.; Brittenham, G.M.; Sroka, R.; Kniebühler, G.; Vogeser, M.; Stepp, H. Bandwidth-variable tunable optical filter unit for illumination and spectral imaging systems using thin-film optical band-pass filters. *Rev. Sci. Instrum.* **2013**, *84*. [CrossRef]

20. Hikmet, R.A.M.; Kemperman, H. Electrically switchable mirrors and optical components made from liquid-crystal gels. *Nature* **1998**, *392*, 476–479. [CrossRef]

21. Li, Y.; Luo, D.; Peng, Z.H. Full-color reflective display based on narrow bandwidth templated cholesteric liquid crystal film. *Opt. Mater. Express* **2017**, *7*, 16–24. [CrossRef]

22. Jeong, M.Y.; Mang, J.Y. Continuously tunable optical notch filter and band-pass filter systems that cover the visible to near-infrared spectral ranges. *Appl. Opt.* **2018**, *57*, 1962–1966. [CrossRef]

23. Lee, K.M.; Tondiglia, V.P.; Lee, T.; Smalyukh, I.I.; White, T.J. Large range electrically-induced reflection notch tuning in polymer stabilized cholesteric liquid crystals. *J. Mater. Chem. C* **2015**, *3*, 8788–8793. [CrossRef]

24. Takezoe, H.; Hashimoto, K.; Ouchi, Y.; Hara, M.; Fukuda, A.; Kuze, E. Experimental study on higher order reflection by monodomain cholesteric liquid crystals. *Mol. Cryst. Liq. Cryst.* **1983**, *101*, 329–340. [CrossRef]

25. Ozaki, R. Simple model for estimating band edge wavelengths of selective reflection from cholesteric liquid crystals for oblique incidence. *Phys. Rev. E* **2019**, *100*. [CrossRef] [PubMed]

26. Chou, S.C.; Cheung, L.; Meyer, R.B. Effect of a magnetic fiedl on the optical transmission in cholesteric liquid crystals. *Solid State Commun.* **1972**, *11*, 977–981. [CrossRef]

27. Dumitrascu, I.; Dumitrascu, L.; Dorohoi, D.O. The influence of the external electric field on the birefringence of nematic liquid crystal layers. *J. Optoelectron. Adv. Mater.* **2006**, *8*, 1028–1032.

28. Jeon, M.Y.; Zhang, J.; Wang, Q.; Chen, Z. High-speed and wide bandwidth Fourier domain mode-locked wavelength swept laser with multiple SOAs. *Opt. Express* **2008**, *16*, 2547–2554. [CrossRef]

29. Ko, M.O.; Kim, S.-J.; Kim, J.-H.; Lee, B.W.; Jeon, M.Y. Dynamic measurement for electric field sensor based on wavelength-swept laser. *Opt. Express* **2014**, *22*, 16139–16147. [CrossRef] [PubMed]

30. Ko, M.O.; Kim, S.-J.; Kim, J.-H.; Jeon, M.Y. In situ observation of dynamic pitch jumps of in-planar cholesteric liquid crystal layers based on wavelength-swept laser. *Opt. Express* **2018**, *26*, 28751–28762. [CrossRef] [PubMed]

31. Jeong, M.-Y.; Wu, J.W. Continuous spatial tuning of laser emissions with tuning resolution less than 1 nm in a wedge cell of dye-doped cholesteric liquid crystals. *Opt. Express* **2010**, *18*, 24221–24228. [CrossRef]

32. Inoue, Y.; Hattori, M.; Kubo, H.; Moritake, H. Faster pitch control of cholesteric liquid crystals. *Jpn. J. Appl. Phys.* **2017**, *56*. [CrossRef]

Article

A Fibre-Optic Platform for Sensing Nitrate Using Conducting Polymers

Soroush Shahnia [1,*], Heike Ebendorff-Heidepriem [2], Drew Evans [3,*] and Shahraam Afshar [1]

1 Laser Physics and Photonic Devices Laboratories, UniSA STEM, University of South Australia, Mawson Lakes, SA 5095, Australia; Shahraam.AfsharVahid@unisa.edu.au
2 ARC Centre of Excellence for Nanoscale Biophotonics, Institute for Photonics and Advanced Sensing, The University of Adelaide, Adelaide, SA 5000, Australia; heike.ebendorff@adelaide.edu.au
3 Future Industries Institute, University of South Australia, Mawson Lakes, SA 5095, Australia
* Correspondence: Soroush.Shahnia@unisa.edu.au (S.S.); Drew.Evans@unisa.edu.au (D.E.)

Abstract: Monitoring nitrate ions is essential in agriculture, food industry, health sector and aquatic ecosystem. We show that a conducting polymer, poly(3,4-ethylenedioxythiophene) (PEDOT), can be used for nitrate sensing through a process in which nitrate ion uptake leads to oxidation of PEDOT and change of its optical properties. In this study, a new platform is developed in which a single-mode fibre coated at the tip with PEDOT is used for nitrate sensing. A crucial step towards this goal is introduction of carbonate exposure to chemically reduced PEDOT to a baseline value. The proposed platform exhibits the change in optical behaviour of the PEDOT layer at the tip of the fibre as it undergoes chemical oxidation and reduction (redox). The change in optical properties due to redox switching varies with the intensity of light back reflected by the fibre coated with PEDOT. The proposed platform during oxidation demonstrates linear response for the uptake of nitrate ions in concentrations ranging between 0.2 and 40 parts per million (ppm), with a regression coefficient $R^2 = 0.97$ and a detection limit of 6.7 ppm. The procedure for redox switching is repeatable as the back reflection light intensity reaches $\pm 1.5\%$ of the initial value after reduction.

Keywords: nitrate sensing; optical fibre; PEDOT; conducting polymer

Citation: Shahnia, S.; Ebendorff-Heidepriem, H.; Evans, D.; Afshar, S. A Fibre-Optic Platform for Sensing Nitrate Using Conducting Polymers. *Sensors* **2021**, *21*, 138. https://doi.org/10.3390/s21010138

Received: 16 November 2020
Accepted: 24 December 2020
Published: 28 December 2020

Publisher's Note: MDPI stays neutral with regard to jurisdictional claims in published maps and institutional affiliations.

1. Introduction

Ions play an important role in many biological and engineering processes. Their concentration in aqueous solution is important for practical applications such as defining the quality of drinking water, the treatment of wastewater and production of food [1,2]. Nitrate is an ion of interest in several of these scenarios; for example, it plays a key role in the nitrogen cycle in agriculture [3], removing the hazardous contaminants from the aquatic environment [4–6], and in measuring nitrate oxide in blood [7]. Quantifying the concentration of nitrate in water is critical to ensure an efficient use of the ions while avoiding any potentially harm impact caused by overuse in environment. Excess nitrate (NO_3^-) from agriculture can ultimately find its way into drinking water due to high water solubility of NO_3^- and be ingested by humans. When in vivo, NO_3^- can be spontaneously reduced to nitrite (NO_2^-), which interferes with the oxygen transport mechanism within the body by the irreversible conversion of haemoglobin to methaemoglobin in red blood cells [8]. To combat this, the Environmental Protection Agency (EPA) has determined the safe levels of nitrogen in potable water to be 10 parts per million (ppm or 10 mg/L) equivalent to 44 ppm of NO_3^- [9,10]. Examples such as this serve as motivation for the specific sensing of NO_3^-.

NO_3^- can be detected through various scientific methods including spectroscopic [11,12], electrochemical [13], chromatography [14,15] and capillary electrophoresis [16]. One of the main accepted detection methods of nitrogen in laboratories is Kjeldahl digestion [17,18]. This method is considered as the reference way of detecting nitrogen in biological samples such as meat, plants and wastewater [19]. In this method, the sample preparation and procedure contains several steps and the detection of other forms of nitrogen containing compounds such as NO_3^- and NO_2^- is not possible without an extra step of reduction to ammonium. Another accepted method by laboratories and scientists is colorimetric detection of NO_3^- by a complex reagent [20–22]. The complex reagent is then reduced to form a coloured compound and the NO_3^- level can be determined by spectroscopy in the visible spectrum. Despite the fact that colorimetric- and spectroscopic-based techniques are highly sensitive (detection levels of only a few ppm), most of them have several sample preparation steps which are time consuming and in many cases hard to apply in the field. Another popular technique is using NO_3^- ion-selective electrodes for deployment in water. The electrochemical performance analysis of NO_3^- electrode has a disadvantage of selectivity. The electrodes typically respond to ions according to Hofmeister selectivity series [23,24]. Recently, the selective uptake of NO_3^- from soil pore water has been demonstrated independent of the Hofmeister selectivity using a thin film coating of poly(3,4-ethylenedioxythiophene)(PEDOT) in an electrochemical process [25]. In a separate study, this conducting polymer material was deposited onto the end of an optical fibre as a step towards a new optical sensor [26].

Herein, the NO_3^- concentration of the order of a few ppm was detected in Milli-Q water using an optical method. An optical fibre coated with PEDOT at its tip is used as the optical sensor. The idea is to take advantage of the small fibre area of exposure and the uptake of NO_3^- by PEDOT and monitor the difference in the back reflection in real-time. In contrast to existing NO_3^- concentration sensing methods with PEDOT [25,27], the current approach does not rely on measuring the potential difference, but rather uses the optical back reflection as a sensing mechanism. Conducting polymers, and especially PEDOT, are well reported to undergo doping and de-doping by inserting and removing anions from within. This reversible process is accompanied by changes in the optical properties of the PEDOT and typically exploited in electrochromic devices [28–30]. For the sensing device reported herein, NO_3^- is used as the dopant, with a basic solution (pH $\approx$ 10.5) of 1 mM CO_3^{2-} used to chemically reduce the PEDOT [31] (similar to hydroxide [32] and polyamide solutions [33]) and thus resetting the sensing probe. The PEDOT doped with the Tosylate anion (PEDOT:Tos) is used to investigate oxidation and reduction (redox) switching in Milli-Q water. The measurement is performed in real-time by monitoring the changes in back reflection from the PEDOT:Tos coated at the tip of the fibre. This study is based on optical reflection at a single wavelength (1550 nm). PEDOT spectra showed higher absorption at longer wavelengths, with the transition of polaron and bipolaron states of charged carriers to a higher order of electron paired and unpaired states that results in a local minima at around 1100 nm [26,34]. This results in a better electro-optical switching at 1550 nm in comparison to visible region and shorter wavelengths [35]. Therefore, a stable and narrow linewidth laser at the telecom wavelength of 1550 nm is a suitable source for this study. Wide availability of fibre-pigtail lasers at this wavelength allows us to have a compact, all-fibre system with no need for optical alignment, thus avoiding power fluctuations. The intrinsic benefits of optical fibres coated with PEDOT, i.e., biocompatibility, small cross section, remote interrogation and widespread communications infrastructure, suggest that the using optical fibre in combination with PEDOT can be considered as a forcible platform for NO_3^- sensing in the future. Both PEDOT and optical fibres, being in vivo friendly [36–39], have a significant edge over other NO_3^- sensing techniques, and the detection of NO_3^- in living organisms is often required in agriculture and health sector. However, the challenge of consistent fibre-PEDOT fabrication and preventing sample contamination still have to be overcome. This work is an initial step in using optical fibre coated with PEDOT for sensing applications.

2. Materials and Methods

The sensing probe consists of a PEDT:Tos layer at the tip of an optical fibre (Figure 1a). The PEDOT:Tos composite was fabricated through a vapour phase polymerization according to a previously published protocol [26]. In summary, this technique involves the following four main steps: (1) Preparation of the oxidant solution. (2) Deposition of an oxidant layer at the tip of the fibre. (3) The fibre coated with oxidant is placed inside a vacuum chamber where it is exposed to monomer vapour. (4) Post-deposition treatment of the PEDOT layer by dipping the tip of the fibre into ethanol to remove any unreacted monomers, oxidants and other by-products. Prior to vapour phase polymerization, an oxidant solution was prepared. The oxidant solutions contain Fe(III) tosylate, which was received from H. C. Starck as a 54 wt.% solution in butanol (Baytron CB 54). Sodium carbonate (Na_2CO_3), sodium nitrate ($NaNO_3$), 3,4-ethylenedi-oxythiophene (EDOT) monomer and the block copolymers poly(ethylene glycol-propylene glycol-ethylene glycol) (PEG-PPG-PEG) (Mw = 5800 Da; trade name = P123) were obtained from Sigma-Aldrich. All chemicals were used without further purification. Various solutions of $NaNO_3$ were prepared in Milli-Q water at different concentrations. A stock solution of 200 ppm (3.23 mM) NO_3^- was prepared. A subsample of this solution was taken and diluted with Milli-Q water to reach 100 ppm NO_3^-. This process of subsampling and diluting was repeated for each new concentration until a series of NO_3^- solutions were prepared: 0.2, 2, 5, 10, 20, 30, 40 and 50 of NO_3^-. The 1 mM Na_2CO_3 solution is used as a basic solution, with a measured pH of 10.5. SMF-28-100 optical fibres having a core diameter of 8.2 μm and a cladding diameter of 125 ± 0.7 μm were obtained from Thorlabs Inc. A Cary 5000 UV–Vis NIR spectrophotometer (Agilent Technologies, CA, USA) was used to conduct the spectroscopy of the PEDOT samples deposited on a soda-lime-silica glass (refractive index, n = 1.5 at wavelength 1550 nm) substrate before and after exposure to NO_3^- and CO_3^{2-} solutions.

Figure 1c illustrates the optical setup for sensing. The light source was a linearly polarized single mode 1550 nm continuous wavelength laser diode beam (Thorlabs DJ532-40, GN5A016). The beam propagates through a coupler; 1% of the optical power measured by a Thorlabs S122C photodiode to probe the stability of the laser during the experiment and 99% of the optical power propagates from terminal 1 to 2 of the circulator. Prior to the experiments, the optical power was set at 2 mW at terminal 2 of the circulator (below the optical damage threshold of PEDOT [26]). The diode laser's output was stable with <0.2% optical power fluctuation over the duration of the experiments. The terminal 2 of the circulator is connected to the input of the sensing fibre. The output end of the sensing fibre was flat-cleaved (≤1°) and coated with the PEDOT:Tos. The coating process has been explained in detail in our previous work [26]. In summary, an oxidizer solution was dip-coated onto the end of the fibre, and subsequently the vapour phase polymerization technique [28] was employed to yield 200–250 nm thick PEDOT:Tos at the fibre end (see Figure 1b). The fibre coated with PEDOT was held stationary to remove any bending loss and/or power fluctuations before dipping into the solution. The solution was centred on a motorized stage and the motor was programmed to bring the solution up and down as the tip of the fibre was dipped into the solution. The light back reflected was passed through terminal 2 to 3 of the circulator and was measured by a Thorlabs S122C photodiode. The difference in back reflected light intensity before and after dipping the sensor in the nitrate solution was used to measure the concentration of NO_3^-.

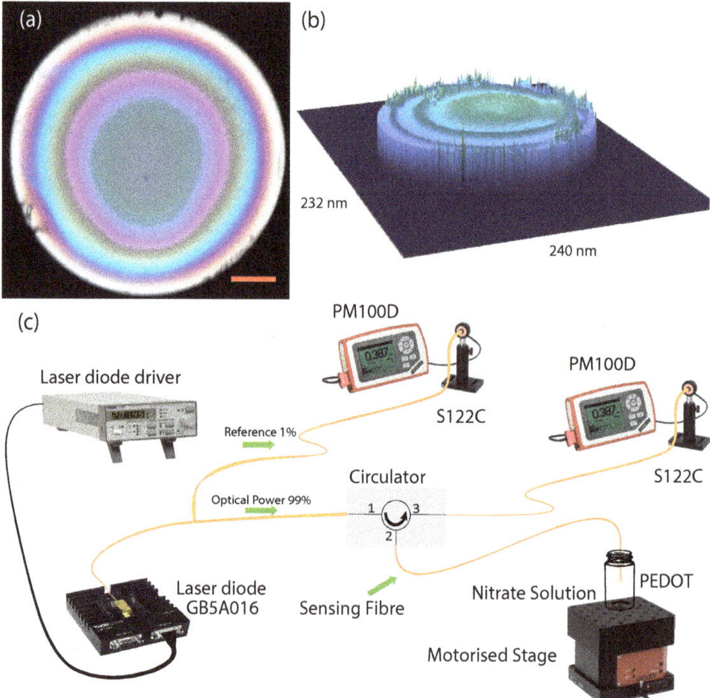

Figure 1. (**a**) Cross-sectional image on reflection mode of PEDOT at the tip of a SMF-28 fibre. Scale bar represent 20 µm (red line), (**b**) 3D surface profile of PEDOT at the tip of the fibre using confocal profilometry and (**c**) experimental setup for measuring NO_3^- concentration based on optical back reflection.

3. Results and Discussion

The spectral response of PEDOT before and after exposure to NO_3^- and CO_3^{2-} has been investigated. In the visible range, the PEDOT layer changes from light blue (as-deposited PEDOT) to a darker blue (after NO_3^- exposure) (see Figure 2a). As the same PEDOT sample is exposed to CO_3^{2-}, it switches to an even darker blue appearance, although the spectral difference is minuscule. The visible change in colour of the PEDOT layer is an indication of the polymers interaction with the surrounding aqueous solution [31]. Figure 2b shows absorbance spectrum (400–2500 nm) of a thin layer (200 nm) of PEDOT prepared on the soda-lime silica glass before and after exposure to NO_3^- and CO_3^{2-} for 2 min. As-deposited PEDOT is highly oxidised, and thus there is little change in the absorption spectra after exposure to NO_3^-. However, there is a decrease in absorption after exposure to CO_3^{2-} due to chemical reduction from the basic solution (pH > 7), in line with previous studies on the pH response of PEDOT [32]. Rather than delve into the mechanism of chemical oxidation and reduction, herein the resultant change in PEDOT optical properties in the short-wave infrared (SWIR) region will be exploited for sensing NO_3^-. This SWIR response conveniently overlaps with the commonly used optical wavelength of 1550 nm (as highlighted in previous studies [26]).

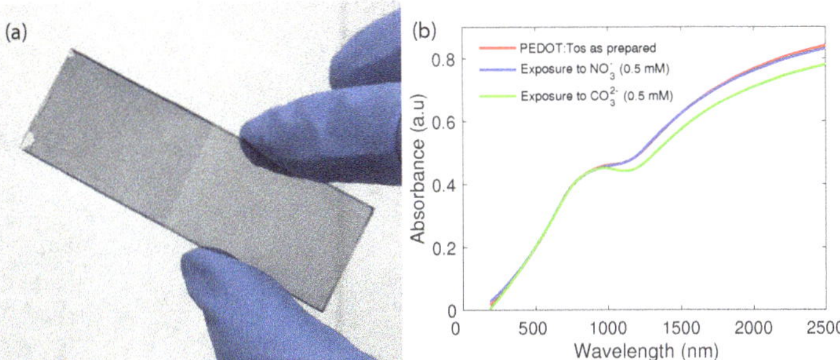

Figure 2. (**a**) The change in PEDOT color from light blue to dark blue in the visible due to oxidation by NO_3^- (1 mM) after 24 h exposure, (**b**) absorbance spectra of PEDOT before and after exposure to NO_3^- and CO_3^{2-} for 2 min.

A different set of experiments was designed to understand the reaction of PEDOT with NO_3^- in Milli-Q water (see Figure 1c). Figure 3a shows the results for two SMF-28 fibres dipped into Milli-Q water: one with PEDOT at the tip, the other a bare flat-cleaved fibre. The tip of both fibres were dipped into a vial containing 15 ml of liquid for a period of 12 h. The magnitude of the back reflection is significantly different when the PEDOT coating is present. The change in back reflection with time is hypothesised to be related to changes in the PEDOT layer thickness, most likely due to the presence of hygroscopic non-ionic triblock copolymer chains (used to template the growth of the PEDOT during the deposition [40]). Long-term soaking of PEDOT in water has been previously reported for the creation of batteries and high humidity sensors [41,42]. After the soaking period, a stock solution of NO_3^- was added dropwise to the vial, increasing its concentration in solution. The increase in back reflection observed for the PEDOT relates to its changing oxidation level as the NO_3^- concentration changes, in contrast to the constant response of the bare fibre (Figure 3b). The constant response for the bare fibre is in agreement with literature studying low concentrations (<1000 ppm) where the change in the solution's refractive index is negligible [43,44]. The change in index of refraction of water is also dependent on temperature and pressure in addition to the NO_3^- concentration [45]. The pressure and temperature in this study are considered constant; atmospheric pressure and the temperature in the optical laboratory are monitored and set at 101.325 kPa and 22 °C, respectively. The results in Figure 3b indicate that the change in the back reflection of PEDOT is due to its oxidation level, not changes in the refractive index of the medium.

As a result of the changing PEDOT oxidation level, the refractive index of the layer also changes, providing a means to optically measure PEDOT's interaction with the surrounding medium. However, when PEDOT is repeatedly exposed to a NO_3^- solution the optical response plateaus (Figure 3c). This plateau occurs when an equilibrium oxidation level is reached for the given NO_3^- concentration being tested/sensed. For example, Figure 3c represents the results where the PEDOT was exposed repeatedly to cycles of 1 min in 50 ppm NO_3^- solution following 1 min in air. The ultimate saturation of the optical response indicates that an additional step is required to realise utility of this sensing platform. The additional step involves exposure to a solution that can reset the oxidation level of PEDOT to a consistent value (and relatively lower compared to those achieved upon NO_3^- exposure).

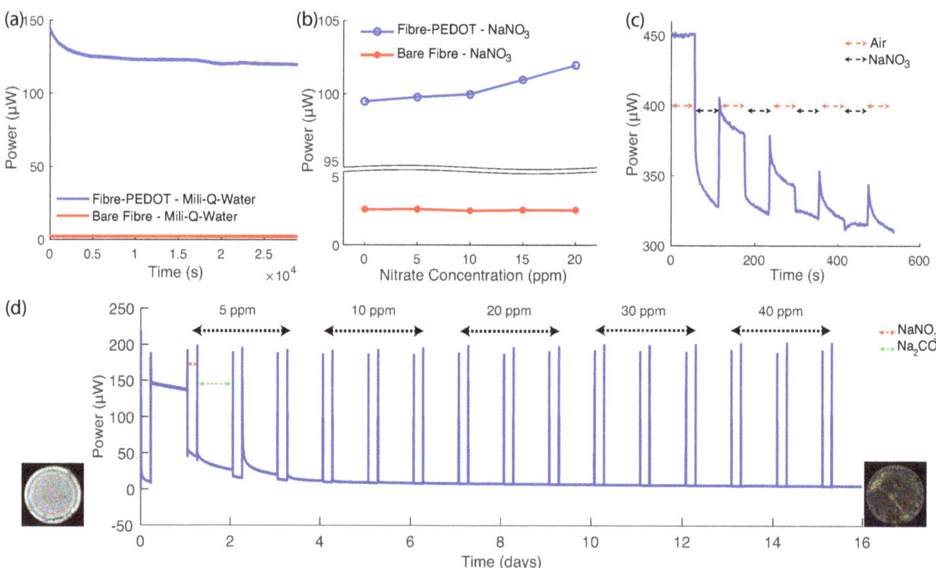

Figure 3. Properties of PEDOT before and after exposure to NO_3^-. (**a**) PEDOT back reflection by reduction in Milli-Q water (blue) and back reflection of fibre (red), (**b**) response of both fibres after addition of droplets of NO_3^- into Milli-Q water, (**c**) a PEDOT sample exposed repeatedly to cycles of NO_3^- solution (50 ppm) and air and (**d**) reusability testing: repeated cycles of $NaNO_3$ and Na_2CO_3 with 2 min of air in between. The image on the left is the cross section image of PEDOT at the tip of the fibre in reflection mode before the experiment and the one on the right is after all the exposure to NO_3^- and CO_3^{2-} for 21 days.

To analyse the optical response of PEDOT at different NO_3^- concentration, a basic solution was employed (pH > 7). Na_2CO_3 was used in solution at low concentration to achieve a pH of $\sim$10.5, in order to chemically reduce the PEDOT to a lower oxidation level. The details of this mechanism and its reversible nature were recently described by Sethumadahavan et al. [31]. In comparison to commonly used NaOH and polyamide solutions, the Na_2CO_3 solution is envisaged to lower the oxidation level of PEDOT without inflicting irreversible damage associated with strongly basic solutions [32]. The change in back reflection upon CO_3^{2-} exposure is correlated with the dynamics of water reacting with the positive charges on PEDOT, causing the liberation of NO_3^-. Figure 3d illustrates the back reflection response of PEDOT at 1550 nm to repeated exposure to $NaNO_3$ and Na_2CO_3 solutions for a range of increasing NO_3^- concentrations (5 to 40 ppm). The sharp peaks in back reflection represent points in time where the fibre is removed from solution to the lab air environment, prior to submersion in the next solution. These peak values correlate well with the concentration of NO_3^- in solution. The before and after exposure microscope images in reflection mode are included in Figure 3d to illustrate what visually happens to PEDOT at the tip of the fibre. PEDOT was initially exposed to NO_3^- for 5 h, then it was dried in air for 2 min. Afterwards, $PEDOT:NO_3^-$ was dipped into the CO_3^{2-} solution for 19 h. The long exposure times is to ensure all the NO_3^- ions are removed from the sample before re-exposing. The sample was dipped into a single NO_3^- concentration 3 times over 3 consecutive days, before proceeding to the next concentration. The back reflection in air (after dip in solution) was compared between the given NO_3^- concentration and the reference CO_3^{2-} solution. The deviation in the response characteristics in back reflection upon exposure to CO_3^{2-} was found to be $\pm$1.5%, highlighting the usability of CO_3^{2-} for PEDOT as a reference point in NO_3^- sensing measurements.

Figure 4 illustrates the difference in back reflection (CO_3^{2-} - NO_3^-) response as a function of NO_3^- concentration. The linear fit shown in Figure 4 produced a regression coefficient of 0.97 with the slope of 0.16 ± 0.03 μW ppm^{-1} for NO_3^- concentrations between

0.2 and 40 ppm. The error bar were calculated by repeating the experiments three times consecutively. The results shows reusability with 21 measurement with one PEDOT sample. The results also allowed for retrieving the limit of detection (LOD), 6.7 ppm, which is defined as LOD = 3σ, with σ = 2.23 being the standard deviation derived from 3 different sample with each one contaminated by NO_3^- and CO_3^{2-} 21 times for a total number of 63 values.

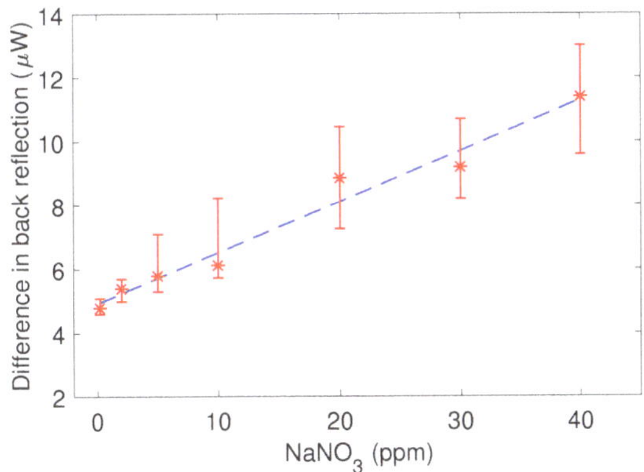

Figure 4. Characterization of the sensor. The difference in back reflection power from the PEDOT layer before and after exposure to nitrate is plotted against the $NaNO_3^-$ concentration. The red stars with error bars represent the experimental data and the blue dashed line represents the linear regression fit with a coefficient of determination of 0.97.

Another important parameter for the sensor characterisation is its response time. Herein, the response time is defined as the exposure time to a particular analyte that is required to achieve a stable back reflection for two consecutive measurements in air after the fibre probe comes out of a particular analyte concentration, while in the above experiments, to make sure that the probe has been fully reset, we immersed the probe in CO_3^{2-} after each NO_3^- exposure for 19 h. In order to see if it would be possible to have a shorten exposure time preliminary series of experiments were conducted. it appears that an initial step of 19 h exposure to CO_3^{2-} to remove most of the tosylate [31] followed by a 2 min consecutive exposure to NO_3^- is sufficient to reset the probe to the same value of back reflection ($\pm1.5\%$). In all measurements, initially fibre probe coated with PEDOT dipped into CO_3^{2-} solution (1 mM) for 19 h as a "resetting" step. In all consecutive steps, the CO_3^{2-} exposure time was reduced to 2 min, as shown in Figure 5a. The probe back reflection has been reset close to its initial value ($\pm1.5\%$) after each CO_3^{2-} exposure which is in agreement with our previous measurements (Figure 3c). After the fibre probe comes out of the Na_2CO_3 and exposed to air for 2 min, it takes less than 4 s for the response time in air to reach the plateau state. Figure 5b, shows the zoomed in results from part Figure 5a for 3 different $NaNO_3$ concentration measurements: 10, 20 and 30 ppm. The time that considered in this experiments is 2 min between each step. This time is required to change the solutions in each step. To check the stability of the sensor, we repeat the procedure 3 times with the same fibre probe and hence the error bar shown in Figure 5b. In this sample, due to lower back reflection relative to the sample in Figure 4, the slope of the fitted line (0.09 ± 0.02 µW ppm^{-1}) is less than our previous measurement. We believe that the main reason for lower back reflection is the variation in sample thickness of ±50 nm and cleave angle of $\pm1°$. However, if we linearly correlated the amount of back reflection to the same level as back reflection as Figure 3c, the slope of the line will be 0.123 ± 0.03 µW ppm^{-1}. The correlated slope is in the range of error and in agreement with our previous results. Therefore, the 2-min interval was considered as the response time.

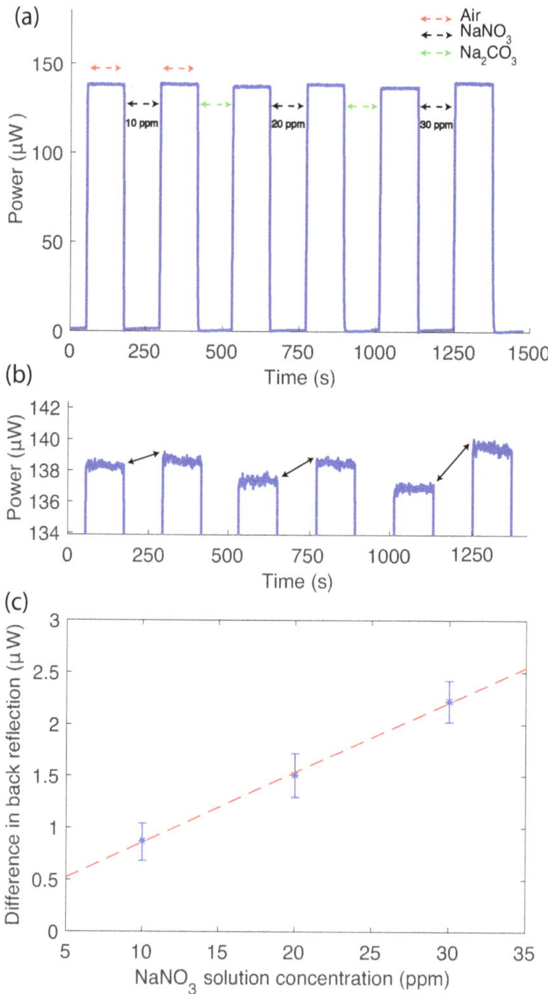

Figure 5. Response time of the sensor. (**a**) Three cycles of NaNO$_3$ and Na$_2$CO$_3$ with 2 min in solution and 2 min in air for different NO$_3^-$ concentration of 10, 20 and 30 ppm, respectively, (**b**) zoomed in section of part a, the black double arrow shows the difference in back reflection before and after exposure to NaNO$_3$, and (**c**) the difference in back reflection light intensity is plotted against the NaNO$_3^-$ concentration. The blue stars with error bars represent the experimental data and the red dashed line represents the linear regression fit.

The conductive polymer PEDOT is observed to repeatedly oxidise with NO$_3^-$ and reduce with CO$_3^{2-}$. The change in intensity of back reflection observed for NO$_3^-$ concentrations as low as 0.2 ppm. Such sensitivity is the point of interest in relation to drinking water, wastewater and water in agricultural soils where NO$_3^-$ is present. The sensor platform surpasses the requirement of measuring NO$_3^-$ concentration below 44 ppm in drinking water by one order of magnitude. Compared to the previously reported work of NO$_3^-$ sensing using PEDOT [25,27], this research employed an optical sensing method rather than an electrochemical one. The current method shows reusability through the ability to "reset" the sensor and this characteristic can be subsequently used for calibration and real-time monitoring. This work provides an alternative approach to some of the conventional and sophisticated NO_3^- sensing methods [17–21]. For example, ion chromatography, one of the most commonly used and accepted techniques, can measure as low as 0.05 ppm but requires an expensive bulky equipment [14,15]. Ramaswami et al. compared a range of techniques to measure NO_3^- in the presence of high concentrations of Cl^- [46]. The reported minimum measurable concentration across those studies was 0.2 and 0.02 ppm. Parveen et al. coated a section of large core fibre (600 μm) with CNTs/Cu-NPs nanocomposite and measured the shift in surface plasma resonance frequency in the visible region of spectrum [47]. They achieved the best value in limit of detection described in the literature, i.e., 0.004 ppm. However, the main issue with surface plasma resonance devices is repeatability. Losses in optical power and alterations in optical coupling drastically changes the performance of such sensors [48]. On top of that, moving the fibre will cause a phase shift, which makes it impractical for applying in the field. In a separate study, an etched fibre Bragg grating was used to measure the NO_3^- by monitoring the shift in the reflected wavelength with a limit of detection of 3 ppm [49]. Fibre-optic sensors offer several advantages over chemical, electrochemical and electric transducers with regards to their small size, in situ/vivo friendly characteristics and the possibility of being deployed as distributed sensors. Here, we developed and studied a chemosensor that has the ability to take advantage of both chemical and optical parameters. From the practical aspect the PEDOT-coated optical fibre herein is easily fabricated and operated, and may overcome some of the practical challenges to allow in-field measurement. Future work will focus on using this platform for the analysis of NO_3^- and NO_2^- compounds in water environments, with particular focus on the ultimate measurement sensitivity, and the detectable concentration range.

4. Conclusions

To conclude, a fibre-optic platform utilising PEDOT at the tip of the fibre has been fabricated and characterised. In this platform, the NO_3^- sensing mechanism relies on redox switching capability of PEDOT. The PEDOT chemically oxidised in the presence of NO_3^- after being reduced within $\pm 1.5\%$ of its initial value in the presence of CO_3^{2-}. The proposed platform during oxidation demonstrates linear response with the slope of 0.16 ± 0.03 μW ppm^{-1} for the uptake of NO_3^- in the concentration range of 0.2 to 40 ppm, with a regression coefficient $R^2 = 0.97$ and a limit of detection of 6.7 ppm. The response time of PEDOT has been measured to be 2 min with the ability of reaching plateau state and recovery in less than 4 s. This new sensing platform is configured as a fibre probe, with benefits such as capability of remote sensing, better area of selection due to a small diameter (125 μm) and less intrusiveness for in vivo measurements. In the future, a fibre-based distributed sensor coated with PEDOT can monitor real-time information on NO_3^- concentrations in a large area such as seawater where collecting discrete samples found to be insufficient for providing statistical significance. Further investigation is in progress to enhance the reproducibility and long term stability of PEDOT, with an ultimate goal of coating on the side of an optical fibre for constructing a distributed optical sensor.

Author Contributions: Conceptualization, D.E., S.A., and S.S.; methodology, D.E., S.A., and S.S.; software, S.S.; validation, D.E., S.A., and S.S.; investigation, S.S.; resources, D.E., S.A., and H.E.-H.; data curation, S.S.; writing—original draft preparation, S.S.; writing—review and editing, D.E., S.A., H.E.-H. and S.S.; visualization, S.S.; supervision, S.A. and D.E.; project administration, S.A.; funding acquisition, D.E., S.A., and H.E.-H. All authors have read and agreed to the published version of the manuscript.

Funding: This research was in-part funded by the Australian Research Council grant number DP170104367 and FT160100300.

Institutional Review Board Statement: Not applicable.

Informed Consent Statement: Not applicable.

Data Availability Statement: The data presented in this study are available on request from the corresponding author. The data are no publicly available as they involve the subsequent application of patent for invention and the publication of project deliverables.

Acknowledgments: This work was performed in part at the South Australian and the Optofab node of the Australian National Fabrication Facility under the National Collaborative Research Infrastructure Strategy. S. Shahnia acknowledges the support of Australian Government through a scholarship under the Australian Research Council discovery project (DP170104367). D. Evans acknowledges the support of the Australian Research Council through FT160100300. The authors acknowledge Saahar Madhavan for his contribution to preparing the sodium carbonate solutions.

Conflicts of Interest: The authors declare no conflict of interest.

References

1. Sethumadhavan, V.; Rudd, S.; Switalska, E.; Zuber, K.; Teasdale, P.; Evans, D. Recent advances in ion sensing with conducting polymers. *BMC Mater.* **2019**, *1*, 4. [CrossRef]
2. Badea, M.; Amine, A.; Palleschi, G.; Moscone, D.; Volpe, G.; Curulli, A. New electrochemical sensors for detection of nitrites and nitrates. *J. Electroanal. Chem.* **2001**, *509*, 66–72. [CrossRef]
3. Alahi, M.E.E.; Xie, L.; Mukhopadhyay, S.; Burkitt, L. A temperature compensated smart nitrate-sensor for agricultural industry. *IEEE Trans. Ind. Electron.* **2017**, *64*, 7333–7341. [CrossRef]
4. Kuddushi, M.; Mata, J.; Malek, N. Self-Sustainable, self-healable, Load Bearable and Moldable stimuli responsive ionogel for the Selective Removal of Anionic Dyes from aqueous medium. *J. Mol. Liq.* **2020**, *298*, 112048. [CrossRef]
5. Mahmud, M.; Ejeian, F.; Azadi, S.; Myers, M.; Pejcic, B.; Abbasi, R.; Razmjou, A.; Asadnia, M. Recent progress in sensing nitrate, nitrite, phosphate, and ammonium in aquatic environment. *Chemosphere* **2020**. [CrossRef]
6. Pięk, M.; Piech, R.; Paczosa-Bator, B. All-solid-state nitrate selective electrode with graphene/tetrathiafulvalene nanocomposite as high redox and double layer capacitance solid contact. *Electrochim. Acta* **2016**, *210*, 407–414. [CrossRef]
7. Carlström, M.; Larsen, F.J.; Nyström, T.; Hezel, M.; Borniquel, S.; Weitzberg, E.; Lundberg, J.O. Dietary inorganic nitrate reverses features of metabolic syndrome in endothelial nitric oxide synthase-deficient mice. *Proc. Natl. Acad. Sci. USA* **2010**, *107*, 17716–17720. [CrossRef]
8. Hilmy, A.; El-Domiaty, N.; Wershana, K. Acute and chronic toxicity of nitrite to Clarias lazera. *Comp. Biochem. Physiol. C Comp. Pharmacol. Toxicol.* **1987**, *86*, 247. [CrossRef]
9. Alahi, M.E.E.; Mukhopadhyay, S.C.; Burkitt, L. Imprinted polymer coated impedimetric nitrate sensor for real-time water quality monitoring. *Sens. Actuators B Chem.* **2018**, *259*, 753–761. [CrossRef]
10. Bendikov, T.A.; Kim, J.; Harmon, T.C. Development and environmental application of a nitrate selective microsensor based on doped polypyrrole films. *Sens. Actuators B Chem.* **2005**, *106*, 512–517.
11. Jahn, B.; Linker, R.; Upadhyaya, S.; Shaviv, A.; Slaughter, D.; Shmulevich, I. Mid-infrared spectroscopic determination of soil nitrate content. *Biosyst. Eng.* **2006**, *94*, 505–515. [CrossRef]
12. Singh, P.; Singh, M.K.; Beg, Y.R.; Nishad, G.R. A review on spectroscopic methods for determination of nitrite and nitrate in environmental samples. *Talanta* **2019**, *191*, 364–381. [CrossRef] [PubMed]
13. Beer, P.D.; Cadman, J. Electrochemical and optical sensing of anions by transition metal based receptors. *Coord. Chem. Rev.* **2000**, *205*, 131–155. [CrossRef]
14. Murray, E.; Roche, P.; Briet, M.; Moore, B.; Morrin, A.; Diamond, D.; Paull, B. Fully automated, low-cost ion chromatography system for in-situ analysis of nitrite and nitrate in natural waters. *Talanta* **2020**. [CrossRef]
15. Wang, H.; Pampati, N.; McCormick, W.M.; Bhattacharyya, L. Protein nitrogen determination by kjeldahl digestion and ion chromatography. *J. Pharm. Sci.* **2016**, *105*, 1851–1857. [CrossRef]
16. da Silva, I.S.; de Araujo, W.R.; Paixao, T.R.; Angnes, L. Direct nitrate sensing in water using an array of copper-microelectrodes from flat flexible cables. *Sens. Actuators B Chem.* **2013**, *188*, 94–98. [CrossRef]
17. Kjeldahl, J. Neue methode zur bestimmung des stickstoffs in organischen körpern. *Z. Anal. Chem.* **1883**, *22*, 366–382. [CrossRef]
18. Muñoz-Huerta, R.F.; Guevara-Gonzalez, R.G.; Contreras-Medina, L.M.; Torres-Pacheco, I.; Prado-Olivarez, J.; Ocampo-Velazquez, R.V. A review of methods for sensing the nitrogen status in plants: advantages, disadvantages and recent advances. *Sensors* **2013**, *13*, 10823–10843. [CrossRef]
19. Sáez-Plaza, P.; Michałowski, T.; Navas, M.J.; Asuero, A.G.; Wybraniec, S. An overview of the Kjeldahl method of nitrogen determination. Part I. Early history, chemistry of the procedure, and titrimetric finish. *Crit. Rev. Anal. Chem.* **2013**, *43*, 178–223. [CrossRef]
20. Shinn, M.B. Colorimetric method for determination of nitrate. *Ind. Eng. Chem. Anal. Ed.* **1941**, *13*, 33–35. [CrossRef]
21. Sancenón, F.; Martínez-Máñez, R.; Soto, J. A selective chromogenic reagent for nitrate. *Angew. Chem. Int. Ed.* **2002**, *41*, 1416–1419. [CrossRef]
22. Khanfar, M.F.; Al-Faqheri, W.; Al-Halhouli, A. Low cost lab on chip for the colorimetric detection of nitrate in mineral water products. *Sensors* **2017**, *17*, 2345. [CrossRef] [PubMed]
23. Hofmeister, F. Zur lehre von der wirkung der salze. *Arch. Für Exp. Pathol. Und Pharmakol.* **1888**, *25*, 1–30. [CrossRef]

24. Andres, L.; Boateng, K.; Borja-Vega, C.; Thomas, E. A review of in-situ and remote sensing technologies to monitor water and sanitation interventions. *Water* **2018**, *10*, 756. [CrossRef]
25. Rudd, S.; Desroches, P.; Switalska, E.; Gardner, E.; Dalton, M.; Buss, P.; Charrault, E.; Evans, D. Relationship between structure/properties of vapour deposited PEDOT and sensitivity to passive nitrate doping. *Sens. Actuators B Chem.* **2019**, *281*, 582–587. [CrossRef]
26. Shahnia, S.; Rehmen, J.; Lancaster, D.G.; Monro, T.M.; Ebendorff-Heidepriem, H.; Evans, D.; Afshar, S. Towards new fiber optic sensors based on the vapor deposited conducting polymer PEDOT: Tos. *Opt. Mater. Express* **2019**, *9*, 4517–4531. [CrossRef]
27. Cheng, Y.H.; Kung, C.W.; Chou, L.Y.; Vittal, R.; Ho, K.C. Poly(3,4-ethylenedioxythiophene)(PEDOT) hollow microflowers and their application for nitrite sensing. *Sens. Actuators B Chem.* **2014**, *192*, 762–768. [CrossRef]
28. Brooke, R.; Cottis, P.; Talemi, P.; Fabretto, M.; Murphy, P.; Evans, D. Recent advances in the synthesis of conducting polymers from the vapour phase. *Prog. Mater. Sci.* **2017**, *86*, 127–146. [CrossRef]
29. Kateb, M.; Ahmadi, V.; Mohseni, M. Fast switching and high contrast electrochromic device based on PEDOT nanotube grown on ZnO nanowires. *Sol. Energy Mater. Sol. Cells* **2013**, *112*, 57–64. [CrossRef]
30. Lock, J.P.; Lutkenhaus, J.L.; Zacharia, N.S.; Im, S.G.; Hammond, P.T.; Gleason, K.K. Electrochemical investigation of PEDOT films deposited via CVD for electrochromic applications. *Synth. Met.* **2007**, *157*, 894–898. [CrossRef]
31. Sethumadhavan, V.; Zuber, K.; Evans, D. Hydrolysis of doped conducting polymers. *Commun. Chem.* **2020**, *3*, 153. [CrossRef]
32. Khan, Z.U.; Bubnova, O.; Jafari, M.J.; Brooke, R.; Liu, X.; Gabrielsson, R.; Ederth, T.; Evans, D.R.; Andreasen, J.W.; Fahlman, M. Acido-basic control of the thermoelectric properties of poly (3, 4-ethylenedioxythiophene) tosylate (PEDOT-Tos) thin films. *J. Mater. Chem. C* **2015**, *3*, 10616–10623. [CrossRef]
33. Bubnova, O.; Khan, Z.U.; Malti, A.; Braun, S.; Fahlman, M.; Berggren, M.; Crispin, X. Optimization of the thermoelectric figure of merit in the conducting polymer poly (3, 4-ethylenedioxythiophene). *Nat. Mater.* **2011**, *10*, 429–433. [CrossRef]
34. Zozoulenko, I.; Singh, A.; Singh, S.K.; Gueskine, V.; Crispin, X.; Berggren, M. Polarons, bipolarons, and absorption spectroscopy of PEDOT. *ACS Appl. Polym. Mater.* **2018**, *1*, 83–94. [CrossRef]
35. Pozo-Gonzalo, C.; Mecerreyes, D.; Pomposo, J.A.; Salsamendi, M.; Marcilla, R.; Grande, H.; Vergaz, R.; Barrios, D.; Sánchez-Pena, J.M. All-plastic electrochromic devices based on PEDOT as switchable optical attenuator in the near IR. *Sol. Energy Mater. Sol. Cells* **2008**, *92*, 101–106. [CrossRef]
36. Luo, S.C.; Mohamed Ali, E.; Tansil, N.C.; Yu, H.H.; Gao, S.; Kantchev, E.A.; Ying, J.Y. Poly(3,4-ethylenedioxythiophene)(PEDOT) nanobiointerfaces: Thin, ultrasmooth, and functionalized PEDOT films with in vitro and in vivo biocompatibility. *Langmuir* **2008**, *24*, 8071–8077. [CrossRef]
37. Venkatraman, S.; Hendricks, J.; King, Z.A.; Sereno, A.J.; Richardson-Burns, S.; Martin, D.; Carmena, J.M. In vitro and in vivo evaluation of PEDOT microelectrodes for neural stimulation and recording. *IEEE Trans. Neural Syst. Rehabil. Eng.* **2011**, *19*, 307–316. [CrossRef]
38. Fani, N.; Hajinasrollah, M.; Asghari Vostikolaee, M.; Baghaban Eslaminejad, M.; Mashhadiabbas, F.; Tongas, N.; Rasoulianborou-jeni, M.; Yadegari, A.; Ede, K.; Tahriri, M. Influence of conductive PEDOT: PSS in a hard tissue scaffold: In vitro and in vivo study. *J. Bioact. Compat. Polym.* **2019**, *34*, 436–441. [CrossRef]
39. Carli, S.; Bianchi, M.; Zucchini, E.; Di Lauro, M.; Prato, M.; Murgia, M.; Fadiga, L.; Biscarini, F. Electrodeposited PEDOT: Nafion composite for neural recording and stimulation. *Adv. Healthc. Mater.* **2019**, *8*, 1900765. [CrossRef]
40. Evans, D.; Fabretto, M.; Mueller, M.; Zuber, K.; Short, R.; Murphy, P. Structure-directed growth of high conductivity PEDOT from liquid-like oxidant layers during vacuum vapor phase polymerization. *J. Mater. Chem.* **2012**, *22*, 14889–14895. [CrossRef]
41. Wang, Z.; Wang, Z.; Zhang, H.; Duan, X.; Xu, J.; Wen, Y. Electrochemical sensing application of poly (acrylic acid modified EDOT-co-EDOT): PSS and its inorganic nanocomposite with high soaking stability, adhesion ability and flexibility. *RSC Adv.* **2015**, *5*, 12237–12247. [CrossRef]
42. Wen, Y.; Xu, J. Scientific Importance of Water-Processable PEDOT–PSS and Preparation, Challenge and New Application in Sensors of Its Film Electrode: A Review. *J. Polym. Sci. Part A Polym. Chem.* **2017**, *55*, 1121–1150. [CrossRef]
43. Esteban, O.; Cruz-Navarrete, M.; González-Cano, A.; Bernabeu, E. Measurement of the degree of salinity of water with a fiber-optic sensor. *Appl. Opt.* **1999**, *38*, 5267–5271. [CrossRef]
44. Jing, J.Y.; Wang, Q.; Wang, B.T. Refractive index sensing characteristics of carbon nanotube-deposited photonic crystal fiber SPR sensor. *Opt. Fiber Technol.* **2018**, *43*, 137–144. [CrossRef]
45. Quan, X.; Fry, E.S. Empirical equation for the index of refraction of seawater. *Appl. Opt.* **1995**, *34*, 3477–3480. [CrossRef]
46. Ramaswami, S.; Gulyas, H.; Behrendt, J.; Otterpohl, R. Measuring nitrate concentration in wastewaters with high chloride content. *Int. J. Environ. Anal. Chem.* **2017**, *97*, 56–70. [CrossRef]
47. Parveen, S.; Pathak, A.; Gupta, B. Fiber optic SPR nanosensor based on synergistic effects of CNT/Cu-nanoparticles composite for ultratrace sensing of nitrate. *Sens. Actuators B Chem.* **2017**, *246*, 910–919. [CrossRef]
48. Allsop, T.; Neal, R. A Review: Evolution and Diversity of Optical Fibre Plasmonic Sensors. *Sensors* **2019**, *19*, 4874. [CrossRef]
49. Pham, T.B.; Bui, H.; Le, H.T.; Pham, V.H. Characteristics of the fiber laser sensor system based on etched-Bragg grating sensing probe for determination of the low nitrate concentration in water. *Sensors* **2017**, *17*, 7. [CrossRef]

Letter

Spectral Reflectance Can Differentiate Tracheal and Esophageal Tissue in the Presence of Bodily Fluids and Soot

David Berard [1,2], Chirantan Sen [1,3], Corinne D. Nawn [1], August N. Blackburn [4], Kathy L. Ryan [1] and Megan B. Blackburn [1,*]

1. Tactical Combat Casualty Care Research Department, US Army Institute of Surgical Research, JBSA Fort Sam Houston, TX 78234, USA; berard64@gmail.com (D.B.); cs3180@msstate.edu (C.S.); nawn.cori@gmail.com (C.D.N.); kathy.l.ryan.civ@mail.mil (K.L.R.)
2. Department of Mechanical Engineering, University of Texas at San Antonio, San Antonio, TX 78249, USA
3. Department of Electrical and Computer Engineering, Mississippi State University, Mississippi State, MS 39762, USA
4. Blackburn Statistics, LLC, San Antonio, TX 78260, USA; august.blackburn.phd@blackburnstatistics.com
* Correspondence: megan.b.blackburn2.civ@mail.mil

Received: 30 September 2020; Accepted: 26 October 2020; Published: 28 October 2020

Abstract: Endotracheal intubation is a common life-saving procedure implemented in emergency care to ensure patient oxygenation, but it is difficult and often performed in suboptimal conditions leading to high rates of patient complications. Undetected misplacement in the esophagus is a preventable complication that can lead to fatalities in 5–10% of patients who undergo emergency intubation. End-tidal carbon dioxide monitoring and other proper placement detection methods are useful, yet the problem of misplacement persists. Our previous work demonstrated the utility of spectral reflectance sensors for differentiating esophageal and tracheal tissues, which can be used to confirm proper endotracheal tube placement. In this study, we examine the effectiveness of spectral characterization in the presence of saline, blood, "vomit", and soot in the trachea. Our results show that spectral properties of the trachea that differentiate it from the esophagus persist in the presence of these substances. This work further confirms the potential usefulness of this novel detection technology in field applications.

Keywords: spectral reflectance; airway management; endotracheal tube placement; tissue detection; tracheal tissue; esophageal tissue; intubation; endotracheal tube misplacement

---

1. Introduction

A fundamental step in front-line emergency medical care is ensuring patient oxygenation. The most common procedure to secure a patient's airway is endotracheal intubation (ETI), in which an endotracheal tube (ETT) is inserted and secured in the trachea. Caregivers face a variety of challenges when attempting ETI, including the difficulty of the procedure itself, environmental variables, patient condition and presentation, as well as obstructive anatomical features of the patient. These challenges increase the rate of failed intubations resulting in patient complications up to and including death [1,2]. ETT placement generally begins when the caregiver visualizes the upper airway using a laryngoscope. However, patient anatomy, injuries such as maxillofacial trauma, swelling, hemorrhage, and other environmental contaminants can increase the difficulty of visualization in as much as 50% of cases in the prehospital setting [3,4]. Due to these difficulties, an array of complications may arise including, hypoxia, hypotension, as well as cardiac arrest resulting from failed intubations [2,5].

Esophageal intubation is one type of failure in ETI in which the ETT is placed in the esophagus rather than the trachea. Habib et al. found this to occur in up to 25% of patients arriving at the emergency department, and if undetected, is fatal [6]. As a precaution against undetected misplacement, the American Heart Association recommends use of a confirmatory procedure following any ETI [7]. Current methods for confirmation include direct visualization of proper ETT placement, which may not be possible due to aforementioned challenges, and the clinically preferred method of end-tidal carbon dioxide monitoring (ETCO2). Although a recent survey revealed that 95% of respondents in the New York State area had capnography and CPR devices aboard ambulances [8], these resources may still not be available in far-forward areas, such as a military setting, and even when available have been found to be underutilized [9]. Additionally, confirmatory methods may require an interruption of patient care or extra equipment that may not be available especially when only a single caregiver is present. Given the time sensitive nature of emergency care, the potential delays and distractions of performing these confirmations can be detrimental to patient outcomes.

Therefore, there is a proven need for an easy, fast, and nonvisual detection mechanism to confirm tracheal placement. We have previously shown that tracheal and esophageal tissues reflect white light differently and produce significantly different spectral profiles in ex vivo and in vivo settings [10,11], as well as in human cadavers [10]. Specifically, tracheal tissue contains a local maximum reflectance intensity at 561 nm with two local minimum intensities at 543 and 578 nm that are not present in esophageal tissues. Thus, the ratios of the reflectance intensities at these key wavelengths, which we refer to as Ratio B (561/543) and Ratio Y (561/578), can differentiate between healthy esophageal and tracheal tissues. We have also demonstrated that these spectral signatures continue to differentiate between esophageal and tracheal tissues in vivo under hypoxic conditions [12]. However, contaminants in the trachea are common in the event of traumatic injury, therefore, it is important to determine if the spectral signatures present in the trachea persist in the presence of contaminants. We hypothesized that spectral reflectance would still differentiate tracheal and esophageal tissues when saline, blood, a simulant of vomit, and soot are present in the trachea.

2. Materials and Methods

2.1. Sample Collection and Spectra Collection

Research was conducted in compliance with the Animal Welfare Act, the implementing Animal Welfare regulations, and the principles of the Guide for the Care and Use of Laboratory Animals, National Research Council. The U.S. Army Institute of Surgical Research Institutional Animal Care and Use Committee approved all research conducted in this study; assigned identification code A-18-012 and approved on 18 January 2018. The facility where this research was conducted is fully accredited by the AAALAC.

The trachea and esophagus were excised from 16 euthanized swine to be studied ex vivo. Once removed, the tissues were stored in a freezer until use and permitted to defrost in a refrigerator for not less than 12 h prior to testing. To collect spectral reflectance data, a custom-designed, 200 μm core, fiber optic probe from Gulf Fiberoptics (R200-7-UV-VIS) was placed into each sample trachea and esophagus (Figure 1). The fiber optic probe contained a centrally located illuminating fiber that carried non-ionizing light emitted from an 8.8 mW LLS-COOL-WHITE halogen light source by Ocean Optics down the 2 m long probe cable where it was redirected 90° by a mirror located at the distal tip of the probe and shown on the adjacent tissue. This design allowed scans to be done with the probe aligned parallel to the tissue and eliminated the need to dissect the tissue structure in order to orient the probe normal to the tissue surface. The probe also contained six signal receiving fibers surrounding the emitting fiber, as shown in Figure 1, that transported the reflected light from the tissue back to a Photon Control SPM-002-DT spectrometer sensitized for 420–1070 nm wave lengths with 2048 × 14 pixel Hamamatsu back-thinned CCD (S9840), and spectral properties were collected from the tracheal and esophageal lumens with the Ocean View Specsoft software package. A "dark" calibration scan was captured after placing the probe in a near perfectly dark environment, and a "light" scan was captured by turning on the light source and placing the probe in a white Ocean Optics FIOS-1 calibration box. Each capture was collected by averaging five subsequent exposures of 5 ms each. Spectral properties of each trachea were measured without contaminants and in the presence of various substances and compared to clean esophagus tissue.

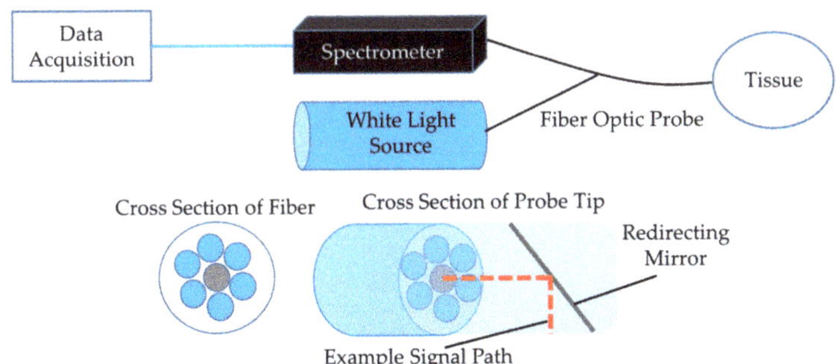

Figure 1. Block diagram of the experimental setup showing data acquisition module and light source connected to fiber optic probe. Cross-section of fiber with emitting and receiving fibers and tip showing mirror with 90° signal redirection included (adapted from previous publication [11]).

2.2. Experimental Group 1

The first set of experiments, $n = 10$, included a 0.9% saline solution, blood (collected during experimental hemorrhage), and simulated "vomit" (cream of mushroom soup), measured in that order. For each swine, three spectral captures were recorded for each contaminant (baseline, saline, blood, and "vomit") in the trachea followed by three spectral captures of the non-contaminated esophagus, resulting in 24 spectral captures per swine in total. Previous work demonstrated that the spectral characteristics of interest were independent of the location within a particular tissue, so each capture was collected at random locations within the treated area [11]. Each substance was introduced to the trachea via syringe and the tissue was thoroughly rinsed and dried between each test. Any sign of tissue injury created during handling, including abrasion, constituted exclusion of that sample. Saline was not captured for one swine and one measurement of "vomit" in trachea was missed in each of two swine. There were a total of 220 spectral captures for this group of swine.

2.3. Experimental Group 2

A second group, $n = 6$, was used to investigate spectral reflectance in the presence of soot in the trachea. Soot was collected by capturing burned paper and reconstituting in saline to make a paste. Figure 2 shows an example of a trachea sample with soot applied. The procedure described above was followed for the application of soot. Twelve spectral captures per swine were taken for a total of 72; to include three baseline measurements in the trachea and esophagus followed by three trachea in the presence of soot and three clean esophagus.

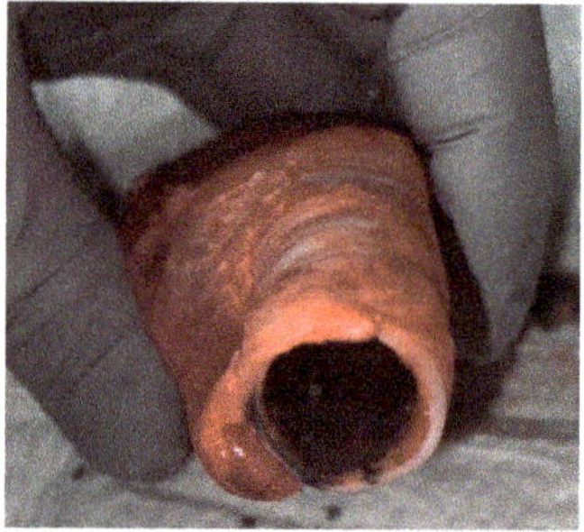

Figure 2. Trachea sample with soot applied prior to spectral collection.

2.4. Spectral Reflectance Processing

The reflectance, absorbance, and amplitude spectra were exported from the Specsoft software as text files and imported into MATLAB for post-analysis. Each capture was identified numerically, according to the experiment, and classified as tracheal or esophageal based on the relevant tissue type. All tracheal and esophageal spectra were first plotted separately to inspect for signal fidelity. Previous work identified 500–650 nm as the range of interest [13], and therefore all spectra were cropped to focus on that region prior to signal processing. Based on previous studies identifying 543, 561, and 578 nm as the wavelengths most useful in distinguishing tracheal and esophageal tissues [13,14], the amplitude values at these wavelengths were extracted for each capture and plotted to visualize the distribution.

2.5. Statistical Analysis

Spectral reflectance was normalized across captures by dividing each spectral reflectance datapoint by the average spectral reflectance across the range of measurements from 500 to 650 nm. This procedure produced an average normalized spectral reflectance of 1 for each capture and retains relative differences between reflectance at different wavelengths. For each capture, the spectral reflectance datapoints were smoothed using kernel regression. Specifically, we used a Gaussian kernel with a bandwidth chosen using the method of Racine and Li [15], implemented within the np package in R [16]. Smoothed reflectance for the wavelengths 543, 561, and 578 nm were extracted for further analysis. Ratio B (561/543) and Ratio Y (561/578) were calculated and used as the primary measures of interest.

The spectral reflectance ratios of interest, Ratio B and Ratio Y, were modelled independently using linear mixed models. The spectral reflectance ratio was treated as the dependent variable. Individuals (swine) were treated as random effects in order to account for non-independence due to multiple measurements per individual. Contaminants (baseline, saline, blood, "vomit", and soot) were treated as random effects and the location of the probe (esophagus vs. trachea) was treated as a fixed effect. Receiver operator characteristic (ROC) curves were used to characterize the diagnostic ability of the spectral reflectance ratios to distinguish esophagus from trachea. Differences in the area under the curve (AUC) between ROC curves were tested using the method described by DeLong et al. [17].

3. Results

Spectral scans of 96 distinct points across 16 clean tracheal and esophageal samples were collected to establish a baseline measurement. At baseline, the characteristic peak at 561 nm along with the adjacent troughs at 543 and 578 nm were present in the trachea while no such features were present in the esophagus, as expected (Figure 3). Tracheal scans taken in the presence of saline, blood, "vomit", and soot continued to show the presence of a spectral peak at 561 nm and accompanying troughs at 543 and 578 nm (Figure 3).

The baseline scans affirmed the results from previous work that identified Ratio B (561/543) and Ratio Y (561/578) as reliable markers to differentiate between tracheal and esophageal tissues (Figure 4).

Tracheal tissue had significantly larger mean Ratio B values for every substance tested when compared to esophageal tissue. The Ratio Y mean was also significantly larger in tracheal than esophageal tissue at baseline and in the presence of saline and "vomit". However, the Ratio Y mean was only nominally larger with soot in the trachea and was smaller with blood in the trachea. These results are reflected in the area under the receiver operator characteristic curves for each ratio, as shown in Table 1 and Figure 5. Overall, Ratio B has a better sensitivity and more specificity in all settings as indicated by the area under the receiver operator characteristic curves (Table 1). This indicates an inability of Ratio Y to distinguish between the two tissues if blood or soot is present.

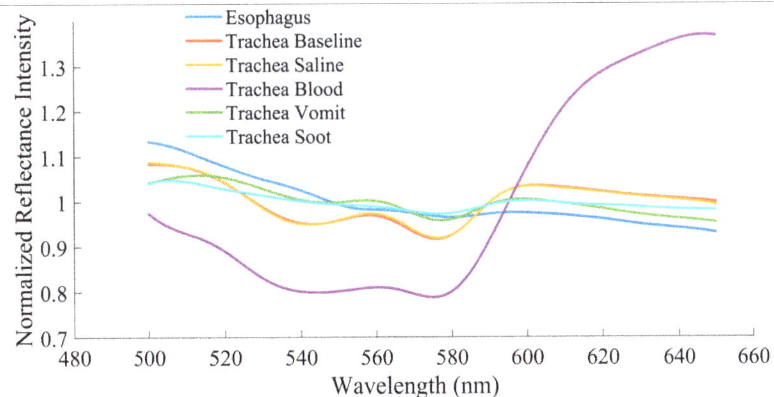

Figure 3. Averaged reflectance spectra at baseline and in the presence of various body fluids for esophageal and tracheal tissues.

Table 1. Area under the receiver operator characteristic curves for differentiating trachea from esophagus using Ratio B and Ratio Y.

Substance	Ratio	
	B	Y
Overall	0.867	0.704
Baseline	0.891	0.871
Saline	0.947	0.878
Blood	0.763	0.364
"Vomit"	0.873	0.801
Soot	0.847	0.557

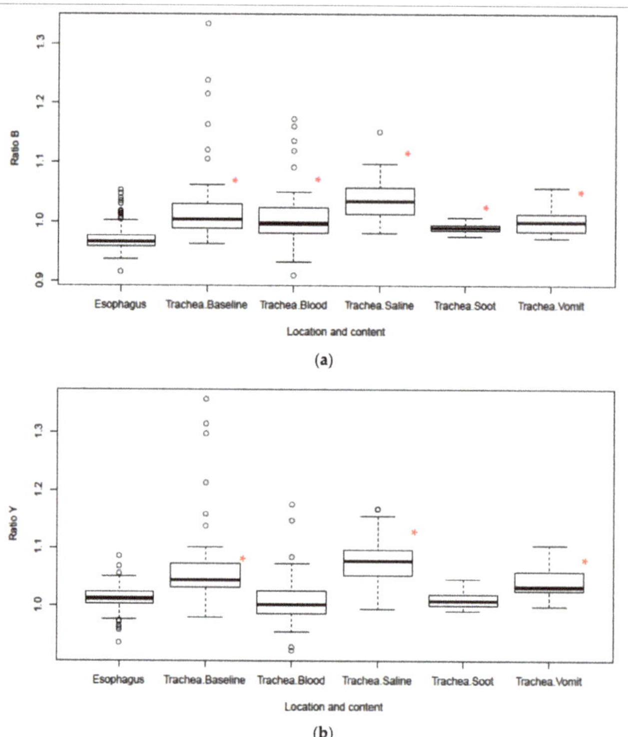

Figure 4. (**a**) Boxplot of Ratio B and (**b**) Ratio Y for tracheal and esophageal tissues in the presence of body fluids and soot. Dots in the boxplots are outliers, defined as being 1.5 times the interquartile range beyond the first or third quartiles. * Represents statistical significance vs. esophagus.

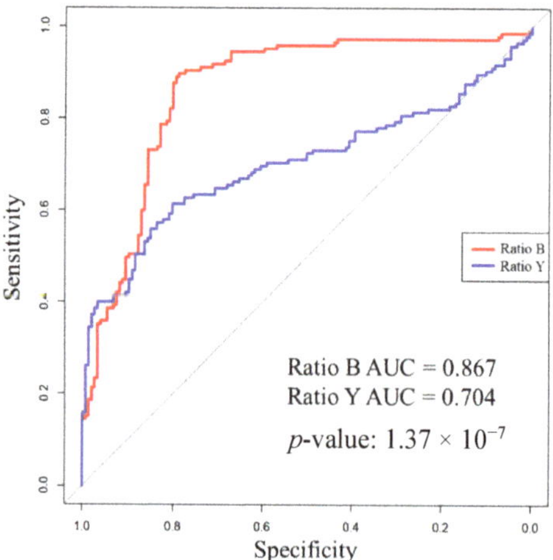

Figure 5. Receiver operator characteristic curves including all Ratio B and Ratio Y measurements in this study with corresponding area under the curve (AUC) values. The AUC for Ratio B is statistically significantly greater than for Ratio Y. In this context, specificity and sensitivity are reported according to correct identification of trachea.

4. Discussion

We evaluated the capabilities of a novel biosensor for detecting proper ETT placement during ETI. Using a fiber optic probe, the spectral reflectance characteristics of tracheal and esophageal tissues can be analyzed and used to differentiate between them, even in the presence of bodily fluids in the trachea which may be encountered in patients with traumatic injury.

Incorporation of a spectral reflectance sensor into the ETT would give providers continuous feedback while performing ETI and provide continuous monitoring to alert providers should the ETT become dislodged during transport. This novel approach shows promise of being more reliable than the current standard of care, ETCO2 monitoring, particularly in pre-hospital settings. ETCO2 monitoring is possible through different techniques including colorimetry, capnography, or capnometry and effectively reduces the amount of ETT misplacements that go undetected [18,19]. The colorimetric approach uses a single-use device housing a chemically treated paper sample. The paper sample reacts when exposed to CO_2 resulting in a color change that can be observed by the caregiver. Unfortunately, the colorimetric method can be slow and subject to user interpretation. Mouth-to-mouth resuscitation, contamination by acidic substances, and various medications can also cause false positive test results [7,20]. The colorimetric sensor is also not capable of providing continuous monitoring, which is problematic as ETTs may become dislodged after initial proper placement [9]. Capnography and capnometry are more quantitative methods that measure CO_2 concentration in exhaled breath [21]. However, the equipment required for these continuous monitoring approaches may be unavailable or underutilized in a prehospital setting. Capnometry in particular relies on higher CO_2 concentrations which may be hindered in instances of cardiac arrest with reduced blood flow and low CO_2 levels [22,23]. If placed into the ETT itself, spectral reflectance sensors may offer better resilience to the aforementioned conditions and provide continuous placement monitoring in places where ETCO2 monitoring is not feasible or unavailable.

Our previous studies identified wavelength ratios B (561/543) and Y (561/578) which can be used to capture the relative prominence of the tracheal peak [10]. However, to date, the ability of these ratios to distinguish between tissue types had not been investigated in the presence of bodily fluids in the trachea likely to be encountered during ETI during traumatic injury, including burn inhalation injury where hot gases and soot particles may be introduced into the airway. Baseline results confirmed previous findings that both ratios were significantly different between the two tissues in an ex vivo swine model [10,11,13]. With saline in the trachea, both ratios remain capable of discriminating between tissue types. This was expected since liquid water is almost perfectly transparent and thus would negligibly affect the reflectance intensities of the tissue. Blood had the greatest impact on the spectra signature characteristic of the trachea, especially with regard to Ratio Y. Figure 3 shows that there is high absorption of wavelengths <600 nm in the presence of blood. Though a local peak and the adjacent troughs are still visually identifiable, they are dramatically blunted in comparison to the tracheal spectra with other substances. This can most likely be attributed to the fact that oxyhemoglobin has relatively high absorption in this wavelength range and low absorption in the 630–700 nm range, which results in its red color [24,25]. Possible higher concentrations of oxyhemoglobin in the tissue layers closer to the surface may also account for the characteristic differences between tracheal and esophageal tissue curves overall [10]. Despite this, Ratio B still differentiated between tracheal and esophageal tissue in the presence of blood, suggesting that its use in a sensor system could overcome this greater absorption. Interestingly, even though tracheal reflectance was impacted by blood, the spectral sensor was capable of differentiating the tissue types in the presence of "vomit" and soot. The relatively opaque substance used for "vomit" and dark color of soot had very little effect on the tracheal reflectance profile.

Overall, Ratio B demonstrated a more robust reliability for distinguishing between tracheal and esophageal tissues than Ratio Y as evidenced by the significantly higher ratios and AUC values for all substances. The overall ROC curves show that Ratio B is significantly closer to the perfect classification, 100% sensitivity (no false negatives) and 100% specificity (no false positive), as compared to Ratio Y. Further studies are warranted to determine whether Ratio B alone would be sufficient or if slightly altered wavelength values might provide greater acuity than the currently selected values for Ratio Y, especially in the presence of blood. Additional research is also needed to determine an optimal clinical strategy for using this information. In the absence of perfect discrimination, which would be reflected by an area under the receiver operator characteristic curve of 1, there are different clinical strategies that should be considered. For example, a threshold for Ratio B could be set to be highly specific or highly sensitive for tracheal placement or to strike a balance between sensitivity and specificity. While we focused on Ratios B and Y in this work, it is possible that other analytical approaches for classifying spectra produced by this device as representing placement in the trachea or esophagus may be superior.

Only four substances were examined in this study with only one substance present at a time. It is also possible that sequentially measuring different contaminants in the same tissue could alter the measurements somehow. We consider this possibility unlikely given the cleaning procedure between contaminants and that the characteristic spectral feature of the trachea persists. Future work may need to include mixtures of these substances as well as others. These four are some of the most likely, but the authors recognize the potential for a multitude of contaminants and their potential effect on spectral characteristics of tissues. Additional studies can also be done on tissue that has undergone trauma since it is currently unknown how certain injuries may impact the reflectance spectra. Inhalation injuries are particularly common in instances of fire emergencies and explosions and often require intubation. It is possible that these effects of inhalation injury on the spectral profile may be quantified as a diagnostic tool for quantifying the severity of the injury. Future studies are also necessary to verify that these spectral properties hold true in living humans. The cost of a clinically used spectral device has not yet been evaluated, however, the available technology can be priced competitively when compared to current small form factor ETCO2 devices. The small footprint and light weight of the components makes this preferable to other larger equipment when these metrics are paramount.

5. Conclusions

We showed that certain ratios of reflectance values at specific wavelengths can be used to distinguish between tracheal and esophageal tissues in an array of conditions likely to be encountered in emergency trauma scenarios. Although the spectral profile of tracheal tissue was altered slightly by the different substances tested, the characteristics distinguishing it from the esophagus remained quantitatively detectable. Ratio B (561/543 nm) proved to be a robust metric that can discriminate tracheal tissue in the presence of saline, blood, "vomit", and soot with reliable AUC values of ROC curves and statistically significant differences in mean ratio values. The absorption properties of blood are most likely responsible for the low reflectance values for wavelengths below 600 nm. In spite of this, some of the qualitative characteristics are still visible, and Ratio B remains significantly different. However, the combination of the general negative slope of esophageal tissue within the area of interest and the reduced intensity of the trachea in the presence of blood and soot caused Ratio Y to no longer reliably differentiate the tissues. Taken together, these results strengthen the potential of spectral reflectance to be used for confirming proper ETT placement in prehospital settings. The probe and accompanying accessories can be integrated into already existing airway devices such as a bougie or ETT.

Author Contributions: Conceptualization, C.D.N. and M.B.B.; methodology, C.D.N., M.B.B.; formal analysis, D.B., C.S., A.N.B.; investigation, D.B., C.S., M.B.B.; resources, M.B.B., K.L.R.; data curation, D.B., C.S., M.B.B., A.N.B.; writing—original draft preparation, D.B.; writing—review and editing, D.B., C.S., C.D.N., A.B., K.L.R., M.B.B.; visualization, D.B., A.B.; supervision, M.B.B.; project administration, K.R.; funding acquisition, M.B.B. All authors have read and agreed to the published version of the manuscript.

Funding: This research was funded by Medical Research and Development Command Tactical Combat Casualty Care Research Program, grant number D_003_2017_USAISR.

Conflicts of Interest: The authors declare no conflict of interest.

DOD Disclaimer: The views expressed in this article are those of the authors and do not reflect the official policy or position of the U.S. Army Medical Department, Department of the Army, DoD, or the U.S. Government.

Animal Statement: Research was conducted in compliance with the Animal Welfare Act, the implementing Animal Welfare regulations, and the principles of the Guide for the Care and Use of Laboratory Animals, National Research Council. The U.S. Army Institute of Surgical Research Institutional Animal Care and Use Committee approved all research conducted in this study; assigned identification code A-18-012 and approved on 18 January 2018. The facility where this research was conducted is fully accredited by the AAALAC.

References

1. Goto, T.; Watase, H.; Morita, H.; Nagai, H.; A Brown, C.; Brown, D.F.; Hasegawa, K. Repeated attempts at tracheal intubation by a single intubator associated with decreased success rates in emergency departments: An analysis of a multicentre prospective observational study. *Emerg. Med. J.* **2015**, *32*, 781–786. [CrossRef] [PubMed]

2. Sakles, J.C.; Chiu, S.; Mosier, J.; Walker, C.; Stolz, U. The importance of first pass success when performing orotracheal intubation in the emergency department. *Acad. Emerg. Med.* **2013**, *20*, 71–78. [CrossRef] [PubMed]

3. Combes, X.; Jabre, P.; Jbeili, C.; Leroux, B.; Bastuji-Garin, S.; Margenet, A.; Adnet, F.; Dhonneur, G. Prehospital standardization of medical airway management: Incidence and risk factors of difficult airway. *Acad. Emerg. Med.* **2006**, *13*, 828–834. [CrossRef] [PubMed]

4. Rudraraju, P.; Eisen, L.A. Analytic Review: Confirmation of Endotracheal Tube Position: A Narrative Review. *J. Intensive Care Med.* **2009**, *24*, 283–292. [CrossRef]

5. Hasegawa, K.; Shigemitsu, K.; Hagiwara, Y.; Chiba, T.; Watase, H.; Brown, C.A.; Brown, D.F. Association between repeated intubation attempts and adverse events in emergency departments: An analysis of a multicenter prospective observational study. *Ann. Emerg. Med.* **2012**, *60*, 749–754.e2. [CrossRef]

6. Habib, N.; Harris, K.; Chalhoub, M.; Maroun, R.; Ciccone, R.; Elsayegh, D. Prolonged Esophageal Intubation: Is it Still Possible? *Clin. Pulm. Med.* **2012**, *19*, 44–46. [CrossRef]

7. Neumar, R.W.; Otto, C.W.; Link, M.S.; Kronick, S.L.; Shuster, M.; Callaway, C.W.; Kudenchuk, P.J.; Ornato, J.P.; McNally, B.; Silvers, S.M.; et al. Part 8: Adult Advanced Cardiovascular Life Support. *Circulation* **2010**, *122*, S729–S767. [CrossRef]

8. Sahyoun, C.; Siliciano, C.; Kessler, D. Use of Capnography and Cardiopulmonary Resuscitation Feedback Devices Among Prehospital Advanced Life Support Providers. *Pediatr. Emerg. Care* **2018**. [CrossRef]

9. Cook, T.M.; Woodall, N.; Harper, J.; Benger, J. Major complications of airway management in the UK: Results of the Fourth National Audit Project of the Royal College of Anaesthetists and the Difficult Airway Society. Part 2: Intensive care and emergency departments†. *BJA Br. J. Anaesth.* **2011**, *106*, 632–642. [CrossRef]

10. Nawn, C.D.; Blackburn, M.B.; De Lorenzo, R.A.; Ryan, K.L. Using spectral reflectance to distinguish between tracheal and oesophageal tissue: Applications for airway management. *Anaesthesia* **2019**, *74*, 340–347. [CrossRef]

11. Nawn, C.D.; Souhan, B.E.; Carter, R.; Kneapler, C.; Fell, N.; Ye, J.Y. Distinguishing tracheal and esophageal tissues with hyperspectral imaging and fiber-optic sensing. *J. Biomed. Opt.* **2016**, *21*, 117004. [CrossRef] [PubMed]

12. Blackburn, M.B.; Nawn, C.D.; Ryan, K.L. Testing of novel spectral device sensor in swine model of airway obstruction. *Physiol. Rep.* **2019**, *7*, e14246. [CrossRef] [PubMed]

13. Nawn, C.D.; Souhan, B.B.; Carter, R.; Kneapler, C.; Fell, N.F.; Ye, J.Y. Spectral characterization of tracheal and esophageal tissues using a hyperspectral camera and fiber optic sensors. In Proceedings of the Optical Fibers and Sensors for Medical Diagnostics and Treatment Applications XVI, San Francisco, CA, USA, 7 March 2016.

14. Souhan, B.E.; Nawn, C.D.; Shmel, R.; Watts, K.L.; Ingold, K.A. Fiber optic tracheal detection device. In Proceedings of the Optical Fibers and Sensors for Medical Diagnostics and Treatment Applications XVII, San Francisco, CA, USA, 28 February 2017.

15. Racine, J.; Li, Q. Nonparametric estimation of regression functions with both categorical and continuous data. *J. Econom.* **2004**, *119*, 99–130. [CrossRef]

16. Tristen, H.; Racine, J. Nonparametric Econometrics: The Np Package. *J. Stat. Softw.* **2008**, *27*, 1–32.

17. DeLong, E.R.; DeLong, D.M.; Clarke-Pearson, D.L. Comparing the areas under two or more correlated receiver operating characteristic curves: A nonparametric approach. *Biometrics* **1988**, *44*, 837–845. [CrossRef] [PubMed]

18. Jones, J.; Murphy, M.P.; Dickson, R.L.; Somerville, G.G.; Brizendine, E.J. Emergency Physician-Verified Out-of-hospital Intubation: Miss Rates by Paramedics. *Acad. Emerg. Med. Off. J. Soc. Acad. Emerg. Med.* **2004**, *11*, 707–709.

19. Silvestri, S.; Ralls, G.A.; Krauss, B.; Thundiyil, J.; Rothrock, S.G.; Senn, A.; Carter, E.; Falk, J. The effectiveness of out-of-hospital use of continuous end-tidal carbon dioxide monitoring on the rate of unrecognized misplaced intubation within a regional emergency medical services system. *Ann. Emerg. Med.* **2005**, *45*, 497–503. [CrossRef]

20. Rabitsch, W.; Nikolic, A.; Schellongowski, P.; Kofler, J.; Kraft, P.; Krenn, C.G.; Staudinger, T.; Locker, G.J.; Knöbl, P.; Hofbauer, R.; et al. Evaluation of an end-tidal portable ETCO2 colorimetric breath indicator (COLIBRI). *Am. J. Emerg. Med.* **2004**, *22*, 4–9. [CrossRef]

21. Berlac, P.; Hyldmo, P.K.; Kongstad, P.; Kurola, J.; Nakstad, A.R.; Sandberg, M. Pre-hospital airway management: Guidelines from a task force from the Scandinavian Society for Anaesthesiology and Intensive Care Medicine. *Acta Anaesthesiol. Scand.* **2008**, *52*, 897–907. [CrossRef]

22. Kodali, B.S.; Urman, R.D. Capnography during cardiopulmonary resuscitation: Current evidence and future directions. *J. Emerg. Trauma Shock.* **2014**, *7*, 332–340.

23. Paal, P.; Herff, H.; Mitterlechner, T.; Von Goedecke, A.; Brugger, H.; Lindner, K.H.; Wenzel, V. Anaesthesia in prehospital emergencies and in the emergency room. *Resuscitation* **2010**, *81*, 148–154. [CrossRef] [PubMed]

24. Nitzan, M.; Romem, A.; Koppel, R. Pulse oximetry: Fundamentals and technology update. *Med. Devices* **2014**, *7*, 231–239. [CrossRef]

25. MacKenzie, L.E.; Harvey, A.R. Oximetry using multispectral imaging: Theory and application. *J. Opt.* **2018**, *20*, 063501. [CrossRef]

Publisher's Note: MDPI stays neutral with regard to jurisdictional claims in published maps and institutional affiliations.